THE AUTHOR

Virginia Paul is married to a cattleman who is a grandson of an early pioneer cattle family.

Mr. and Mrs. Paul live on a small ranch East of Ellensburg where they raised four children, three sons and a daughter.

She has spent the last fifteen years in service to the agricultural community through her association with state farm and ranch organizations.

The areas of her responsibilities have been varied and have ranged from office management to the development and direction of education, promotion and public relations programs.

She is enthusiastic and this is reflected in her dealings with people and the imaginative ways she handles materials for publication. Mrs. Paul has helped substantially in the preparation of publicity for statewide use in all newsmedia and has contributed regularly to periodicals circulated to members of the organizations for whom she has worked.

These include the Washington State Farm Bureau Federation, The Washington State Cattlemen's Association, the Washington Beef Council, the United Good Neighbors of Kittitas County and the Washington State Beef Commission. She has served as Administrative Secretary for the Washington State Beef Commission since its creation by Legislative Act in June, 1969.

She is a member of the Evergreen Chapter of the American Women in Radio and TV, the Washington Chapters of the American National CowBelles and National Press Women. Virginia was recognized by the latter group in 1970 for "outstanding achievement and community service in the field of communication."

She has received state and national awards in the category of radio script writing for her beef education program which is featured regularly on sixteen stations in Washington state.

is was
ttle Ranching

"Circle Diamond" Cattle near Hinsdale. John Survant's outfit, foreman, Bloom Cattle Company, Trinidad, Colorado.

Courtesy Montana Historical Society, Helena, Montana

This was Cattle Ranching
Yesterday and Today

BY VIRGINIA PAUL

Superior PUBLISHING COMPANY

708 SIXTH AVE. NORTH, SEATTLE, WASH.

FIRST EDITION

DEDICATION

**DEDICATED TO THE NORTHWEST
CATTLE INDUSTRY**

TABLE OF CONTENTS

FOREWORD

The history of the cattle industry is unexcelled for drama and excitement.

Almost every child rides a stick horse and drives an imaginary herd of wild, longhorn steers across sand and sage.

Even when adulthood is reached and making a living reality, it is refreshing to succumb to the magic lore of cowboys, cattle drives and cattle barons.

Such imagery is an integral part of Northwestern heritage in fact as well as fiction.

It is not the intent of this book to disillusion or detract from this remarkable image but to broaden its scope with authenticity.

Long ago, cattle ranching was a part of everyone's life. In current times, few people have the opportunity of personal acquaintanceship with the industry.

The book is about the real world of cattlemen and cowboys; of cattle drives and drovers and of an industry that is unique. Why does the cattle industrialist risk so much for so little?

To learn the reason and to record a fascinating history, I have traveled to all parts of the Northwest and spent three years gathering information and photographs.

I have received the wholehearted cooperation of many pioneering families. The candid language of such cattle folk became the basis for much of the narrative. In searching their collections of family pictures and snapshots, they recalled, as accurately as possible, dates, names and events.

Among the photographs are those of Huffman, Barker and others, who, without the aid of sophisticated camera equipment, captured the drama of yesterday's ranching scene.

Many scenes are from fifty to seventy years old. A few were photographed more than one hundred years ago and, though faded and worn, were included to help "tell it like it was."

The early cattle ranching industry was a "man's world." Today, it is the rare ranch woman who longs for further liberation. Many share equally in the daily work of making a living on a ranch. Often, a rancher's wife is his "top hand," bookkeeper, and partner in addition to homemaker.

There is no way to "get rich quick" in modern beef business. "You have to marry it or inherit it," a rancher will tell you with a grin.

After embarking on a livestock venture, no well-coordinated, overall marketing concept ensures a consistent return on a ranch investment, partly because of a cattleman's independent nature.

Historically, prices fluctuate from year to year. Periods of higher return do not fully compensate for years of lower prices. The price paid for the beef product is not based on a cost-plus factor "You take what you can get when the beeves are ready to go."

"This was Cattle Ranching" covers more than one hundred years of cattle ranching history, each chapter of which could be of book length.

Because there are limits to time, opportunity and printed space, many individuals who have contributed much to the development of the industry have not been included.

I hope readers of all ages will enjoy adding the real story of Northwest cattle ranching to the glamor of the past and will gain a better understanding of what it is from the story of what it was.

Many fine individuals helped make this book possible. My deep appreciation to those credited within and to the following:

Mrs. Walser S. Greathouse, Mrs. John Harting, Queena Davison Miller, Ilda May Hayes, John Quigley, Joseph Schnebly, Dick Blue, John Armstrong, Tim Eckert, Irene Stangoe, Dave Hamm, Charles Jarrett, Joe W. Jarvis, Homer Splawn, Roger Roberts, Port Griswold and Frank Mitchell for resource information;

Mickey Moe, Grace Weber and Mike Steigleder for photographic processing;

The various cattle breed associations of national and international scope for providing breed histories;

The cattlemen's associations of British Columbia, Idaho, Oregon, Montana and Washington for referrals to pioneer cattle families;

The county and state historical societies for invaluable additions to the book content;

The directors and staffs of libraries and archives for willing and capable research assistance;

The departments of agriculture and brand registrars for providing records and documents;

R.L. "Buzz" Capener for inspiration;

Emerson Matson for encouragement;

Al Salisbury for patience, guidance and the opportunity.

CHAPTER I
Since Time Began

From prehistoric to modern times, cattle have influenced man's destiny and will continue to affect his economic, physical and social environment. And from the discovery of fire to the conquering of space, the most significant contribution made by the cattle family has been that of helping man maintain a more satisfying level of nutrition and living.

Enlightening discoveries made in layers of earth dating back to the Stone Age, link the Giant Ox (Bostaurus primigenious) of ancient Europe to the much smaller Celtic Ox (Bos longifrons). An ancient relationship is recognized between the huge, wild Asian ox, humped Bibovines, various buffaloes, the Bison and cattle proper.

Whatever the origin or relationship, this predominant species furnished power, provided hides for covering and supplied milk and meat for human sustenance since the beginning of agriculture. These benefits to mankind are so appreciated and respected the animal is held sacred in some lands still. In many primitive and advanced countries around the world, the larger the herd, the more esteemed the owner.

Through the ages, herdsmanship became a more exacting and sophisticated science. By improving methods in care, feeding and breeding, the animal was adapted to better suit specific needs.

Of necessity, early cattle in the Northwest served multi-purposes but as human and cattle populations increased more attention was devoted to the development of breeds for milk or for meat.

To increase beef production, much success has been achieved by visionary individuals through good utilization and management of hardy beef stock and lands rich with natural resources.

These changes have not come easily nor always kept abreast of the need. Progress has been accomplished by sacrifice and sometimes with avarice, but never without great perseverance, untiring vigor, high expectations, poignant drama, keen humor and tense excitement.

Of such is the substance of the Northwest cattle industry and the people who have contributed to its fascinating history and growth.

Today, raising beef is a highly diversified and scientific business, contributing millions of dollars to the economy of America's Big Northwest Country.

Old Bos
The ancient Teutons called this animal the Aurochs. The coat of long hair was dark brown to black and fossilized skeletons indicate the animal was about six feet in height at the shoulder.
The animal ranged over the greater part of Europe. Modern cattle are believed to have originated from the Aurochs and the Celtic Shorthorn.

Buffalo grazing in the Flathead Valley, Montana, on the Moiese Buffalo Preserve.

The Buffalo (Bison) was the native cattle of the North American Continent. A civilization depended upon this animal for food, clothing, shelter and fuel. The buffalo was of such significance that the early American's life-style and religion were based on the animal's existence.

Courtesy Montana Historical Society, Helena, Montana

Zebu

Cattle have served many purposes. The Indian Zebu (Brahman) has been used for centuries in the Orient to till the soil and to produce milk and meat.

Certain religious sects, among them the Hindu consider cattle sacred and the meat is not consumed.

Courtesy Northwest Farm Unit Magazine.

Bull team

Pulling together by eight or a dozen, bull teams demonstrated their ability and strength to move mountains of timber, haul freight and other heavy loads in the early 1900's.

World's largest ox team

Tom and Jerry, the world's largest oxen, weighing approximately 4,000 pounds each, were raised and driven about the turn of the century. This team broke all records for pulling and was widely known at the time. The animals were purebred Holsteins but were unregistered since oxen may not be considered breeding animals.

Courtesy Northwest Unit Farm Magazines

Ezra Meeker, an early settler of the Northwest, retraced his journey of previous years to the West, Meeker located in the Puyallup Valley of Washington. Photo taken in 1906.
Courtesy of the Oregon Historical Society, Portland, Ore.

Eight oxen pulling a boiler in 1886 from Pomeroy, Washington to the Andy Hert sawmill 15 miles south of Pomeroy, an uphill pull.
Courtesy Robert R. Beale.

Teams of oxen once hauled freight outfits up the Cariboo Road. Shown is one of the early day outfits in front of the old Clinton Hotel which burned to the ground in 1958.

Early pioneers used from one to three yoke of oxen for each wagon. Oxen were slower than horses but superior in stamina.

Courtesy of Williams Lake Tribune, B.C.

Wild Range Cattle
Double wintered Montana-Texans.

The influence of English breeds on Texas Longhorns can be noted in the white faces of the cattle as a start is made to increase meat-producing ability.

Courtesy Montana Historical Society, Helena, Montana.

CHAPTER II
Era of Expansion

The Pacific Northwest Region of the North American continent was one of the last frontiers to be settled. The American Colonies had declared their independence from England before any permanent settlements of consequence had been established in this sector of the New World.

The area comprising the Province of British Columbia, Canada, and the present states of Washington, Oregon, Idaho, and Montana is a land of geographical variety and unsurpassed scenic beauty — high mountain ranges, rushing rivers, jewel-like lakes, grassy plateaus and arid desert land. The total acreage in round figures is 482,000,000, an area larger than one third of the United States.

In the Western region, rainfall is released in mountain ridges reducing moisture in the Eastern portion.

The Canadian Province and states have a common watershed, the largest component of which is the Columbia River and its tributaries.

An abundance of natural resources contribute to the appeal of the Northwest — forest and wildlife, water, valuable ores and fertile soils.

The area was opened to settlement from the north by English interests in their search for furs and the subsequent establishment of the Hudson Bay Fur Trading Posts in the late 18th and early 19th Centuries. From the south, Spanish explorers sailed north along the Pacific Coast claiming land for the King of Spain.

Between 1803 and 1805, under authorization by the United States Government, Lewis and Clark formed an expedition, explored and charted the inland territory below the 49th parallel. The explorers followed the Missouri River westward to its source, then continued on a hazardous, painstaking, arduous investigation of the streams and rivers flowing to the Pacific Ocean.

Nothing less than a holocaust of great expanse and long duration would have deterred the settlement of the Northwest following this blazing of trails.

In the middle of the 19th Century, gold was discovered and, with its discovery, a mass of humanity poured into the area from all directions. Those infected with this "Fever" abandoned local enterprises and traveled by boat, foot, horseback, packmule and wagon train to the streams and rivers in every state and British Columbia.

Hauling supplies to remote sites, which grew in a short period of weeks from a few tents or shacks to cities of from three to ten thousand or more, was a gargantuan task. Most of the supplies were brought by pack animal at great difficulty through rough terrains; were sold at good profit by the supplier and purchased dearly by the miners and local merchants. Among the main sources of supply were Victoria, Portland, The Dalles, Umatilla and Walla Walla.

Better means of transportation contributed further to the occupation and settlement of the Northwest Territory. Nearly twenty-seven years after the steamboat was invented, the "Beaver" was leaving a tumbled wake in the waters of Puget Sound and the mouth of the Columbia. Within a few years, other steamboats were employed to navigate rivers far inland, providing a fairly dependable service except during severe winter months.

In 1883 the Northern Pacific Railway linked the East to West followed by the Canadian Pacific in 1885 and the Great Northern two years later. The Southern Pacific was completed from California to Oregon in 1887 joining the territory to the south. These railroads plus other lines of less magnitude, furnished faster passenger and mail service and added greatly to the amount of freight that could be hauled over distances separating the populace from its source of supply.

Fort Vancouver
An artist's conception of Fort Vancouver as it was in 1854.
Courtesy Oregon Historical Society, Portland, Oregon.

Steamboats by river, improved trails, wagon roads and railroads by land, lessened the economic risk and burden of survival for early pioneers. Better transportation encouraged the imigration of more and more families seeking broader horizons and greater potential for economic security.

Though some were opportunists, most were hardy, courageous, sincere people of various ethnic, cultural and social backgrounds, looking forward to building a new life.

The building of Northwest cattle industry closely parallels the foregoing historical summary of development with a few exceptions peculiar to cattle raising.

The first cattle in the Northwest were of Spanish origin, dating back to the 16th Century when Spanish explorers and Conquistadors brought cattle to the southern part of the North American Continent. Two centuries later small numbers of Spanish cattle had been moved as far north as Nootka Sound. The Northwest cattle industry began with this nucleus.

Very little occurred in the way of herd improvement and growth until Dr. John McLoughlin came to the Columbia District as chief factor for the Hudson Bay Fur Trading Company. A need existed for posts to be self-sufficient because of distance and terrain. Such independence was encouraged.

Dr. McLoughlin consolidated cattle at Vancouver moving some from the abandoned post of John Jacob Astor's Pacific Fur Company. He imported

Dr. John McLoughlin, Chief Factor of the Hudson Bay Fur Company for the Columbia District.
Courtesy Oregon Historical Society, Portland, Oregon.

15

John Ferrell's pack outfit going to cowcamp in the Tieton Basin. Pack trains traveled over narrow Indian trails laden with goods and supplies for remote settlements and mining camps. Pack animals were used before roads were improved enough to permit travel by freight wagons.

Courtesy Julia Ferrell Hopf

several Durham cattle and a few sheep from England. As much as he felt it desirable to upgrade Spanish cows for increased milk production, importing improved cattle in quantity from England was not feasible.

He realized the agricultural potential of the district because of soil and climate. From a business viewpoint he saw the production of food and fiber as a valuable addition to fur trade.

Assisted by friendly Indians, with whom Hudson Bay had been conducting a profitable and fairly harmonious fur business, Dr. McLoughlin cultivated the land and planted crops. And cattle multiplied for he allowed none to be used for meat but kept for milk and reproduction purposes.

Though herds continued to thrive, Dr. McLoughlin zealously sustained Hudson Bay interests by refusing to sell livestock to American settlers. Realizing the privations of pioneer life, however, his humanitarian character prevailed and he began a practice of loaning two cows each to settlers. He also encouraged and materially assisted the crop and livestock pursuits of other Hudson Bay posts.

Before long, livestock and by-products were in such quantity that Dr. McLoughlin felt the Company's interests could best be served if the agriculture

Handmade wagon wheels used in the construction of Fort Colville.

Photo by author.

16

investments were a separate entity. The Puget Sound Livestock Company was formed under the auspices of the Hudson Bay Company with headquarters at Fort Nisqually. This Company was purchased by the Americans after the treaty and agreement on boundaries.

Thus, the cattle industry was launched as a major enterprise in the Northwest.

The discovery of gold provided the impetus for sufficient import of cattle to satisfy the hearty appetites of miners and inhabitants of booming gold rush towns, and to better supply the needs of growing settlement. Cattle drives from California to Oregon, from Texas to Montana and inter-territorial movements among Oregon, Washington, Idaho, Montana and British Columbia began a cattle trade that is unprecedented in American history for drama and excitement. This commerce created the picture of the Western Cowboy as a footloose, hardworking, harder playing, earthy-talking, colorful, skillful man with a horse and saddle, a gun and lariat.

Spending months on the trail over rocky mountain terrain, and crossing broad, grassy prairies, and shale-covered deserts, with protection from the weather a rare privilege, was a task only the most hardy or desperate of men undertook. In addition, skirmishes with Indian tribal members were not uncommon.

Kamiakin was one of the first chiefs in the Oregon Territory to foresee a shortage of wildlife as a source of food for his people. In the 1840's he traveled to Fort Vancouver where he secured cattle from Dr. McLoughlin to build a domestic meat supply. These cattle plus the tired, worn-out cattle of immigrants coming West by wagon train for which the Indians traded horses, formed the nucleus of Indian cattle herds. Within five or six years, cattle represented a major share of Indian wealth. Following a massacre at the Whitman Mission in 1847, and subsequent warfare throughout the territory, the Indians became impoverished as American troops confiscated and slaughtered herds to force submission. After Indian reservations were established, cattle raising was slowly resumed.

Carrie Ladd

Stern wheelers such as the "Carrie Ladd" transported passengers, supplies and livestock on Northwest waterways. The Williams Lake Tribune, British Columbia, reported in the early 1870's "the Sir James Douglas arrived in Victoria from Nanaimo bringing a few passengers, 20 hogs and six head of cattle. The passage down was unpleasant." The cost of freighting goods from the mouth of the Skeena River to the Omeneco gold fields was 2½¢ per pound.

In 1851, horses cost $6.00 per head to ship from The Dalles to Portland, Oregon. Cattle boats were used for cattle shipment out of Puget Sound to Victoria and from Oregon to coastal settlements to the north. The average cargo consisted of 30-40 head of live cattle and sheep, dairy products and salted or brined beef.

Courtesy Oregon Historical Society, Portland, Oregon.

S.S.R.P. Rithet in 1862
Bringing fortune seekers by the thousands and supplies were small "river paddlewheelers" such as the S.S.R.P. Rithet shown moored at the boisterous city of Yale, B.C.
    Yale lies north of Hope on a rocky ledge of the Fraser River and was reputedly one of the roughest, most lawless gold mining towns in the Northwest in 1858.
Courtesy Provincial Archives, Victoria, B.C.

The Indians realized their land was being irrevocably claimed. They lived in fear of hunger as they watched the extermination of the buffalo, deer and other native meat animals. They visualized the great risk of survival presented by increasing numbers of people with superior weapons of defense.

Neither Indians nor settlers had an insatiable desire to subjugate, but by accident or design, by provocation or desperation, each became a serious threat to the other. Of little credit to humanity, both committed atrocities.

Pioneers were harassed in traveling and at settlements. Cattle herds were raided on ranch and trail. Retaliation was common and domination inevitable.

As more land was brought under cultivation and barbed wire facilitated the installation of fences to protect crops, traffic in cattle was seriously impeded.

The Great Northwest's empire of grass was diminishing as cattle herds grew in number and as settlers plowed under lush, native stands for the growing of grain and other field crops.

Occasionally, severe winters reduced cattle numbers by the thousands, leaving grazing lands covered by skeletons; reducing men of wealth to poverty status within days.

Gradually the industry changed from that of roving cattle herds with unhindered, unbounded grazing privileges to the limitations of individual ranches where hay could be harvested and stored and more attention given to herd and range improvement.

Further encroachments have been placed upon agricultural lands by the demands of an expanding, mobile population. Year by year, increasing numbers of cattle are confined to less and less earth, contributing to the ever-present challenge of converting feed into beef.

Wright's Ranch, 127 Mile House at Lac La Hache, British Columbia. Freight teams pulling out of the station at the turn of the 19th Century.

Courtesy Provincial Archives, Victoria, B.C.

Cattle herded through the streets of Barkerville, B.C., a roaring mining town which gradually dwindled to a few people by the end of the century. It boomed once again with the discovery of gold in 1920. The picture was taken of cattle for miners being driven down the main street after the fire of September 1868.

Courtesy Provincial Archives, Victoria, B.C.

Thirty years later, herding cattle through Barkerville on the way to Bold Mountain for summer pasture in 1898.

Courtesy Provincial Archives, Victoria, B.C.

Freight wagons pulled by five mule teams along the Thompson River at Grants Bluff, eighty-eight miles above Yale, B.C. in 1868.

Courtesy Provincial Archives, Victoria, B.C.

Cattle drive of Texas Longhorns from a drawing by Edward Borein.

Courtesy Montana Historical Society, Helena, Montana.

Northern Pacific Railroad Stock Train in the 1890's.

Courtesy Montana Historical Society, Helena, Montana

Having a place to rest was a blessing in the days when traveling was rough and slow. The Barbour House at Pemberton, B.C. was such a haven as early as 1908. It was leveled in 1950. Photo taken in 1947.

Courtesy Provincial Archives, Victoria, B.C.

A common sight in 1915 was this 26-horse team with three jerk lines.

Two stages of Wyers Stage Line meet on the old Oakridge Road on the way to Glenwood, Washington.

The area around Glenwood has been cattle country since the early days.

The stage line was owned and operated by Teunis Wyers from 1894 to 1963. The line was motorized in 1922 but horses continued to be used during the winter months until the early 1930's. Circa 1916.

Courtesy Mr. and Mrs. Russell R. Kreps.

CHAPTER III
Starting From Scratch

Knowing the route west was hazardous, uncomfortable and slow, deterred few men and families looking forward to establishing a cattle ranch in the Northwest.

Susan Brewster, a young lady of 17, traveling with her Uncle Ed (who planned to raise cattle), kept a diary of wagon caravan travel to the Frontier. The caravan headed west from Ohio in the spring of 1851 and arrived at The Dalles, Oregon in September.

In April's cold, stormy weather, the wagon train started out. After traveling five miles, camp was made on the "open prairie in the wind and the rain."

The next day Susan noted that "nothing particular happened worthy of note" although one yoke of oxen had drifted away delaying the train until two o'clock in the afternoon. She commented, "everything went pleasant and agreeable. Not much swearing among the boys."

The caravan continued Westward and "crossed Rowan's Ford on the North River (Nebraska) where they found plenty of corn and hay for the cattle."

From time to time, delays were caused by straying oxen and injured stock. Major, the horse pulling Susan's carriage, "got hurt through the night on a stub so we found he would not be able to go at all. We were obliged to put on a yoke of oxen before the carriage and leave Major," Susan lamented.

Before coming to the Sweete River (Wyoming), the wagon train covered many miles of very sandy roads through rough, sagebrush land.

Susan was excited about her first sight of Independence Rock (Wyoming), an immense, flat-topped mass rising to one hundred feet in height.

Thousands of names were written or engraved upon its surface, evidence of the number of immigrants who had passed it on their way to California gold fields and Oregon Territory.

The seventeen year-old secretly noted: "We did not engrave our names but left a note on one of the highest points in a crevice of the rocks for Mr. William's Company of Illinois as he had a number of nice young men with him."

During June, travel was more hazardous and uncertain. Melting snows turned streams into rushing rivers.

All able members of the caravan helped make rough bridges where streams were too swollen and swift to attempt fording the rushing water.

On a warm, pleasant day in July, after passing through a canyon in high mountains, the wagon train came in sight of the Great Salt Lake City, Utah which at first "looked like an old ploughed field," Susan reported. "But as we neared it, it began to look better. The streets are very regularly laid out. They were very wide and a beautiful stream of water runs through every street. The City is four miles square. The dwellings are mostly small, made of doby."

Susan and her uncle traveled seven miles farther and camped where they intended to stay for a few days. Her uncle's plans were to sell merchandise which he had brought across the plains in wagons. The money from sales was used to purchase cattle in the Salt Lake City area.

A few days stretched into twenty before the transactions were completed, the cattle herded together, camp broken and their journey continued.

By now, the heat was intense and the land through which they traveled barren. The dust rose in clouds. Dead cattle marked the way.

Susan's compassion for the people who had traveled the road before her mounted as she saw abandoned possessions scattered along the trail. "We saw a great many dead cattle and wagons, that and other property that had been left by the poor immigrants, where their cattle had given out until they were obliged to leave everything."

Ten oxen pulling logs on skids to the Andy Hert Sawmill, Garfield County, Washington, in 1887. These oxen had been used to cross the plains.

Courtesy Robert R. Beale

Thus, the caravan arrived at Rocke Creek (Wyoming) where the road crossed in a narrow canyon. The party assembled on a high bluff to spend the night before traveling on.

"We supt on butter milk, Indian pudding and went to bed with spirits revived."

Before coming to the Lewis River (Idaho), the caravan passed a large village of Indians "who came running with their arms full of things to swap."

After the long and arduous journey, Susan's joy at reaching The Dalles, where she was reunited with a young man "whom she feared was dead," was beyond written expression. Briefly, she mentioned arrival at The Dalles, "the end of the rainbow," and an end to an early chapter of her life.

Susan married C.P. Cooke the young man who met her. They became one of the first families to settle in Central Washington where they started a ranching enterprise carried on today by descendants.

Not all pioneer cattlemen had family ties. Many were responsible to no one, eager to test and taste life to the fullest and ready to tackle any occupation. Among business prospects to be found in the booming western frontier were trading goods, freighting, blacksmithing, logging, cattledroving, mining, farm

work and cowboying, all of which could lead to something bigger and better.

Jason Lee, an early traveler across the plains who was filled with zeal to convert Indians to Christianity, was the first to trail a few head of cows and a bull over the 2,200 mile-long Oregon Trail in the 1840's. After developing a Methodist Mission at The Dalles, he became interested in farming.

"Uncle Dan" Drumheller was just a lad of eleven or twelve when he became excited about the discovery of gold in California. He received permission from his mother to accompany Joe McMinn and his family from St. Louis, Missouri to a new life in the Golden Territory. McMinn planned to take 175 cattle to California where the market was $90 a head, a gross return of $70 each.

McMinn had gathered together a party of several families which included six men, three women and some children. There were three wagons drawn by four yoke of oxen each, plus six head of horses for handling the herd of cattle. Young Dan was allowed to join the party because of his "way" with cattle and horses.

Upon reaching California he hired out to a cattleman, and within a short period of time entered the

The site of 150 Mile House, British Columbia, was originally owned by the Davison Brothers in 1861 who were engaged in farming. The brothers sold to A.S. Bates who developed a hotel, store, blacksmith shop and ran 400 head of cattle.

Gavin Hamilton, Chief Factor with the Hudson Bay Company, bought the enterprise from Bates in 1876 for $35,000 and sold it ten years later to Veith and Borland at a substantial loss.

The partners rebuilt the business to its former prosperity, selling it 15 years later to an English concern for $90,000. Mrs. Hugh Cornwall, whose father managed the ranch and later purchased the entire property, remembers hearing as a child, "that in the gold rush days, the 150 Mile was known as the "banana belt" of the Cariboo. Many miners from Barkerville, Horsefly and other mining areas, spent their winters there and turned it into a pretty lively place."

Her father ran the ranch in connection with his Onward Ranch, another very old place. The property carried more than one thousand head of mixed-breed cattle, Mrs. Cornwall recalls "Over the years my father bought good Hereford bulls and culled the herd. When we sold the property in 1950, it was running 700 head of top quality Herefords."

Courtesy Williams Lake Tribune

Northwest where he bought, sold, and trailed cattle throughout Oregon, Montana, Idaho, Washington and British Columbia.

His brothers, Tom and Jesse, had preceeded him to the Western settlements. Tom became a placer miner around Hangtown and Red Dog, California and finally farmed in Washington State. Jesse settled a few miles south of Walla Walla in 1855 and, when it was found that wheat would grow on the grassy rolling hills, concentrated on the raising of grain.

The intrigue of the open country drew merchant John R. Quigley to Montana Territory a decade before the Battle at Little Big Horn. He opened stores in Virginia City and Diamond City, later moving his business to another booming, mining town called Blackfoot City. Having a keen mind for business, he encouraged his sons to develop cattle ranches in the Avon-Helmsville area.

Cattleman Marcellus "Ted" Bascum Hayes was born in 1864 on the Tommy Edward donation claim near Springfield, Oregon. He grew up in Sycon Country, alert in his boyhood for Indian signal fires on the hills. When 23, he traveled to Harney County. Having no funds, he recalled earning "his first four-bits catching a horse for Warren Jordan." He started a herd of Devons, switching to Whiteface Herefords, as many early cattlemen did.

The earliest settlers selected the choice spots near rivers, springs, and lakes to homestead. Later arrivals were forced to claim less productive land but made up for quality in vast holdings of free range land, an arrangement well-suited to cattle raising.

Putting up living quarters was secondary to claiming the land. Rough shelters were built quickly and simply of sod, rocks, logs or unplaned lumber. Roofs were sometimes topped with soil and grass and it was a rare home that had a board floor.

However good or poor the lodging, neighbors, travelers, drovers and cowboys were received warmly. Indians were frequent callers and were accepted with some reservation because Indians did not understand the "Boston" (Chinook for American) or his attitude in regard to personal possessions. It was a common occurrence for one Indian, or more, to walk into a home and take whatever was handy. On the other hand, Indians were known to respond by sharing their own food with settlers in need.

During those early years, self-reliance and fortitude were absolutely essential to personal survival and success in cattle ranching.

In the 1870's and 80's the typical ranch consisted of 160 acres of homestead land plus the surrounding free range government land. The herd was composed of from two to five thousand head although there were a few cattlemen who had herds four or five times as large.

As the mining markets for beef tapered off, cattlemen found themselves with a surplus and a falling market.

The decline in price was climaxed by disasterous winters in the late Nineteenth Century which few cattle survived, forcing many cattlemen out of business. The land of opportunity was still there but it had changed.

Cattlemen continuing in the cattle ranching business installed more fences, increased feeding and rebuilt their herds.

Men such as these, wealthy or not, prominent or unnoticed — whatever their origin and wherever located — made cattle ranching what it is. They contributed to the growth and future of the Northwest, helping turn Indian trails into roads, shacks into homes, packtrains into railroad trains, wilderness into settlement and an opportunity into an industry.

Pierre Wibaux was the owner of one of the largest herds of cattle of all times, more than 100,000 head. His cattle brand was "W Bar". He brought many steers from Texas and was a banker in Miles City, Montana in later years. Wibaux, Montana bears his name for it was from this area that he launched a most successful career. He was born in 1856.

Courtesy Montana Historical Society, Helena, Montana

Pierre Wibaux's first Montana home.

Courtesy Montana Historical Society, Helena, Montana

Residence and office of Pierre Wibaux, Wibaux, Montana.

Courtesy Montana Historical Society, Helena, Montana

Contract for steers entered into in April 12, 1894 for delivery in May.

Courtesy Robert R. Beale

THIS AGREEMENT, entered into this ....12"....day of....april....Town....189.4 by and between............
....John Beck....party of the first part, and....E Burlingame....
Pomeroy, Garfield County, Washington, party of the second part: WITNESSETH, That said party of the first part hereby sells to the party of the second part the following described cattle, to-wit:....Seventeen head of Steers....

....6...Steers one year old $ 8 00/100....per head........per pound........

....8...Steers two years old $ 13 00/100....per head........per pound........

....3...Steers three years old $ 18 00/100....per head........per pound........

........Steers four years old $........per head........per pound........

All of said cattle are to be in good shipping condition, full age, good size, free from bulls, stags, big-jaws, cripples, the same to be from my herd branded thus:
....CI on hip....., the said cattle to be delivered by the party of the first part, in good condition, to the party of the second part, at....Rummess Coral....on or about the ....1st....day of....May....189.4
and (....J....) hereby acknowledge receipt of....Twenty and 00/100....Dollars, as part payment of same; the balance to be paid at time of delivery, when accepted by the party of the second part.

John Beck [SEAL]

WJ Rummers....Witness.   E Burlingame [SEAL]

John S. Devine was a pioneer Harney County rancher in Oregon. He built up extensive property holdings which he sold to Miller and Lux, owners of a large cattle operation in the Northwest.
Courtesy Oregon Historical Society, Portland, Oregon

Granville Stewart did many things and was actively interested in the cattle industry. During the 1880's, with Davis and Hauser, both men of means, he put together a cattle empire in Central Montana, the DHS outfit. He served the Montana livestock industry as president of the livestock association and the Board of Stock Commissioners. In late life he became interested in politics.
Courtesy Montana Historical Society, Helena, Montana

Conrad Kohrs started with a butcher shop and ended owning an empire of cattle.

With a partner, Ben Peel, he opened a meat shop on Last Chance Gulch (later the main street of Helena). The partners purchased the Racetrack Ranch in Deer Lodge Valley and cattle at St. Ignatus, Bitteroot and around Fort Benton. Johnny Grant's ranch and cattle were added to the holdings. Kohrs anticipated cattle market trends and, for many years, influenced the cattle trade in Western Montana Territory.

Courtesy Montana Historical Society, Helena, Montana

Thomas S. Blythe was born in Scotland July 5, 1853. He was known as Lord Blythe after coming to the United States and finally settling in Washington State. From 1885 to 1906 he operated a large cattle ranch in Grant County. He owned 3500 head of Shorthorn cattle branded with the "Railroad Brand," two long, parallel bars down the left flank. He could be easily set apart from the cattlemen of that time because of his trim goatee and monocle.
Courtesy Tom Drumheller, Jr.

Home of Lord Blythe

Sam T. Hauser was a Helena Montana banker, financing many cattle for Granville Stewart and other early cattlemen.
Courtesy Montana Historical Society, Helena, Montana

Moreton Frewen was one of the opportunists who saw the potential offered by the great Northwest land of grass. He was an Englishman and married to an aunt of Winston Churchill.

His greatest asset was his ability to gain financial backing for cattle enterprises which included large holdings in Wyoming and Montana, and later in British Columbia.

Frewen lived extravagantly and entertained lavishly. His business ventures were a succession of failures. Contributing to his loss in range cattle was the severe winter the latter part of the 1880's, which ended the cattle careers of many other men as well.
Courtesy Montana Historical Society, Helena, Montana

Tom and George Drumheller The Drumheller Brothers (Tom, left) bought out Lord Blythe in 1906. The brothers ranged a herd of 3,000 Herefords and eight bands of sheep from the time they bought the place until 1938.

Both were wheatmen in Walla Walla with interests in retail hardware and racehorses.

The Drumheller Brothers were well-known for their bucking stock and wild horses which they furnished to Rodeos including Walla Walla, Pendleton, Cheyenne and Calgary.
Courtesy Tom Drumheller, Jr.

Herman Oliver, one of the early pioneer ranchers of Oregon. Oliver settled in Harney County where he developed a solid ranching business. He contributed to the growth of his community and Oregon's beef industry.

Mr. and Mrs. Charles Moon and Violet.
Courtesy R.A. Moon, Williams Lake, B.C.

The Deer Park Ranch, named for thousands of deer that grazed on the native pasture-land, was originally settled by a bachelor in 1860 and consisted of 160 acres of land. It was purchased by Charles Moon and developed into 2,000 acres of deeded land plus 2,000 acres of permit grazing land. The ranch has carried 275-300 head of cattle for many years. Deer Park Ranch is located near Williams Lake, B.C. Cattle are branded with the Maltese Cross Bar brand. Picture taken in 1901.

Courtesy R.A. Moon,
Williams Lake, B.C.

John Walls Snook and John Wilson Snook, father and son
John W. Snook (right) was the oldest living pioneer (1971) in Lemhi County, Idaho. He freighted and hauled supplies for gold rush towns just as his father did. He was a State Legislator in 1909, a warden at the Boise Penitentiary for ten years and at Atlanta, Georgia for five. Snook served as U.S. Territory Deputy for Alaska where he met his wife. When he was eight years old he adopted the Half Circle S brand, one of the oldest brands in Idaho. He was born in 1876.

Courtesy John W. Snook.

The Snook home built in the late 1890's. The ranch was a part of the present site of Salmon, Idaho.
Courtesy John W. Snook

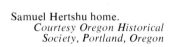

Samuel Hertshu home.
Courtesy Oregon Historical Society, Portland, Oregon

Nelson Story home at Bozeman,
Montana
*Courtesy Montana Historical
Society, Helena, Montana*

The "White House." Pete
French ranch house.
*Courtesy Oregon Historical
Society, Portland, Oregon*

Clarence Dooly, pioneer Washington Rancher.

Old Figure 3 Homestead (more than 100 years old) Clarence Dooly Ranch, Sprague, Washington.
Courtesy Mr. and Mrs. Clarence Dooly

Old Spear Cow Camp on Bitter Creek. "Bitter Creek Flows into Powder River not far from the Montana-Wyoming line. The ranch house stands a couple of miles from the river and was still intact in recent years. It was the "SA" Cow Camp. In addition to a bunch of Spear Children, it sheltered many a trail driver and not a few outlaws." L.A. Huffman photograph.
Courtesy Montana Historical Society, Helena, Montana

Mackay-Fulton Ranch house. Circa 1890. Roscoe, Montana area.
Courtesy Montana Historical Society, Helena, Montana

A hardy Scot, Joseph Dixon Lauder, owned and operated a ranch near Douglas Lake when range land was still open and free. Fourth generation Lauders still ride the family range.

Courtesy Provincial Archives, Victoria, B.C.

A view of the BX Ranch residence (F.S. Barnard) site near Vernon, B.C.

Courtesy Provincial Archives, Victoria, B.C.

Johnny Grant at 36 years of age. Johnny was born in 1831 at Fort Des Prairie (now Edmonton, Alberta). He settled permanently at Deer Lodge, Montana in 1859. He built the first house in Deer Lodge later owned by Conrad Kohrs.

Courtesy Montana Historical Society, Helena, Montana

Close up shows home detail with Frank Barnard and Gertrude Oliver (Mrs. Herman Robertson) getting ready to enjoy a hunt by horseback. Circa 1898.

Courtesy Provincial Archives, Victoria, B.C.

William J. Roper of Cherry Point.
Once a sailor, then a freighter on the Cariboo Road, he started a cattle herd on Cherry Creek which he developed into a sizeable outfit by 1886. He used the Anchor Brand.
Courtesy Provincial Archives, Victoria, B.C.

"Bill" Roper's spread was at Cherry Creek around 1887. He knew many tales of old trail men he had heard as a lad. The 108 Mile House was a place where weary travelers were welcome. Circa 1887.
Courtesy Provincial Archives, Victoria, B.C.

Interior of a ranch house on the Powder River, Montana. L. A. Huffman photograph. *Courtesy Montana Historical Society, Helena, Montana*

Interior of the Australian Ranch home (British Columbia) photographed in 1867. Seated by the fireplace are Andrew (left) and Sam Olsen.
Courtesy Provincial Archives, Victoria, B.C.

Johnny Dick homestead near Peola, 1900. Mr. and Mrs. Dick, Al, Effie and Olive Dick. Al Dick became a prominent Southeastern Washington cattleman.
Courtesy R.P. Weatherly.

Fred G. Little of Creston, B.C.
Courtesy Provincial Archives, Victoria, B.C.

F.G. Little's home site was cleared out of the wilderness. Willow Grove Ranch as it appeared in the early days. Circa late 1800's.
Courtesy Provincial Archives, Victoria, B.C.

Ranch home in Madison County, Montana.

Courtesy Montana Historical Society, Helena, Montana

Original home of the Sinlahekin, the first ranch in Okanogan County, Washington. It was once the headquarters of Phelps and Wadleigh, early cattle buyers and drovers, sold to Guy Waring in 1875 and purchased by John Woodard.

Courtesy Ross Woodard

Sinlahekin Ranch in 1930.

Mr. and Mrs. William Pinch-beck. At the time of his death in 1893, Pinchbeck owned the entire valley where the town of Williams Lake, B.C., now stands. Part of the property included the old Borland House near the lake. Pinchbeck came to the area in the 1860's.
Courtesy Williams Lake Tribune.

Views of the Pinchbeck ranch home and buildings in 1887.
Courtesy Provincial Archives, Victoria, B.C.

Homeranch headquarters of the Bow and Arrow brand, organized in 1885. The picture was taken in 1893, showing Coal Creek emptying into Sunday Creek. The wagon boss, Guy W. Whitbeck, wagon and tent are shown in the immediate center. Rea Cattle Company, Miles City, Montana with ranges on Sunday Creek, North and South Forks Creek, Muster, Harris and Custer Creek.

Courtesy Montana Historical Society, Helena, Montana

George Bosley farm at Columbia Center, ten miles south of Pomeroy, Washington. The farm is completely gone. Picture taken in 1900.

Courtesy Robert R. Beale

Fall ends and winter begins at the Miller farm at Pemberton, B.C.
Courtesy Provincial Archives, Victoria, B.C.

Archdale Ranch of M.M. Arch-
dale, Knowlton, Montana.
*Courtesy Montana Historical
Society, Helena, Montana*

Al Dick ranch headquarters
near Peola, Washington in 1920.
Courtesy R.P. Weatherly

The John Manly Ranch at
Grand Forks, B.C. then owned
by Martin Burrell.
*Courtesy Provincial Archives,
Victoria, B.C.*

Ranch site in the Quesnel District of British Columbia,
fenced with slender poles.

Stump Lake Ranch near Kamloops, British Columbia. It was here that Jim Frisken decided to start a small ranch as headquarters for horsetrading. Frisken came from Scotland in 1883 to demonstrate farm machinery manufactured in his native land. Following this endeavor he worked for a short time at railroad construction in the Rocky Mountains.
Courtesy Provincial Archives, Victoria, B.C.

Eastern Oregon Ranch in 1937. Federal Writers Project photograph.
Oregon Historical Society, Portland, Oregon.

Steve Black ranch headquarters on Smoothing Iron Ridge, Asotin County, Washington, 1938.

Courtesy R.P. Weatherly

Beautiful ranch settings in the Northwest are more common than rare. A view of Joseph Creek Ranch, belonging to Jack Tippett, 1958.

Ranch headquarters in Nicola Valley, Sherman County, Oregon.

Courtesy Oregon Historical Society, Portland, Oregon

50

Paul Bretesche ranch scene in the late 1880's at Trail Creek, Carbon County, Montana.
Courtesy Montana Historical Society, Helena, Montana

At the Corral on Trail Creek. 1880 Paul Bretesche photograph.
Courtesy Montana Historical Society, Helena, Montana

Corral scene near Stinking Hot Water Springs by Paul Bretesche.
Courtesy Montana Historical Society, Helena, Montana

The John Ranayne Farm at Pemberton, B.C. in 1911.
Courtesy Provincial Archives, Victoria, B.C.

An early ranching scene along the Cariboo Road.

Courtesy Provincial Archives, Victoria, B.C.

Australian Ranch at Quesnel, B.C. in 1905.
Courtesy Provincial Archives, Victoria, B.C.

Ranch scene south of Fort Ellis, Montana. W.A. Jackson photograph.
Courtesy Montana Historical Society, Helena, Montana

Putting bells on cows at the Al Dick Ranch in 1940. Ranch headquarters near Peola, Washington.

54

With young grandson, Harold Wick, hitching a ride, as Floyd M. Bradbury harrows the family garden plot at Challis, Idaho in 1939.

Courtesy Lawrence F. Bradbury

Hauling manure on The Ranch (Cotton) around 1912, Williams Lake, B.C.

Courtesy Mrs. C. Moon

Ross Woodard and "Mick" Kinchelo with army remounts sold in 1939. Okanogan County, Washington.

Courtesy Ross Woodard

Chores. Evelyn Jephson Cameron milking, probably on the E.S. Cameron ranch near Terry, Montana.

Courtesy Montana Historical Society, Helena, Montana

AX Ranch of Joe Axford on the Deschutes River in 1945. Al Monner photograph.
Courtesy Oregon Historical Society, Portland, Oregon

Gifford Photograph. Courtesy Oregon Historical Society, Portland, Oregon

A cattle ranch scene of 1940 taken at the Al Dick Ranch, Peola, Washington.

Jack Tooles Willow Creek Ranch in the Sweet Grass Hills of Montana in 1960.
Courtesy Montana Historical Society, Helena, Montana

At the home corral. R.A. "Pudge," Violet, J.C. and Melville Moon (Dorothy not shown). Practicing with the lariat is a favorite pastime of all ranch youngsters and once acquired, a practical skill.

Courtesy R.A. Moon, Williams Lake, B.C.

Coleman

Woodard

Dooly

"Cattlemen never retire—they just saddle up and ride on."

Courtesy R. A. Moon, Williams Lake, B.C.

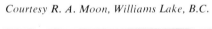

McRae

Snook

58

CHAPTER IV
Cattle Drives and Drovers

In 1836 Ewing Young led the way to the most glamorous and exciting commerce explosion in the history of Western development when he drove the first herd of Spanish cattle from California to the Northwest Territory. Though his immediate goal was to help satisfy the needs of pioneer settlement, the then bold venturer blazed a trail of importation and proved to future generations of ambitious men that cattle could be moved on the hoof in large numbers in spite of the hostile nature of man, weather and land.

Names will be carried forward in history of the dynamic men who gathered cattle together and moved them from Texas to Montana, from Washington and Oregon to Idaho and Montana, from Montana to British Columbia, from lower British Columbia to the Yukon and from interterritorial points to the Eastern markets — names such as Nelson Story, Ben Snipes, Jack Splawn, the men of the XIT* and Matador — among the first to bring the most.

These men shared a dream, to capitalize on the opportunity offered by the discovery of gold, growing settlement and the broad, grass covered areas waiting to be turned into meat.

Early cattle drovers traveled light, packing supplies on horses or mules and sometimes traveling without benefit of pack animals — just man, horse, bedroll and a sack of beef jerky. Later, in trailing larger herds of 2500 or more head, a trail boss supervised the cattle and cowboys, about six riders to every thousand head of cattle. Also included in the drive were six or more horses for every rider, cared for by a "horse wrangler." Wherever and whenever terrain permitted a chuck wagon and cook were employed, which eased the life of the cowboy.

*The XIT and Matador were large cattle outfits in Southwestern United States. The XIT brand meant "ten counties in Texas."

Travel was slow. A dozen miles was considered a good day's journey. Starting at day break, herds grazed for several hours on the move. At noon a halt was called to allow the drovers to eat and to rest the cattle. In the afternoon two or three hours more were devoted to grazing along the way and then men and cattle "bedded down" for the night. The site selected was called "bed-ground."

"Nighthawkers," generally working in pairs, were assigned to "night herd." The position was shared alternately by all members of the crew and was not met with enthusiasm in freezing rain. Nor was there a clamor for the chore of 'drag' — following along behind cattle on the move in dust too thick to breathe when it was dry or in trampled, sloppy mud when it was wet.

More experienced cowboys and the "trail boss" rode in lead positions.

Between 1859 and 1870, more than 22,000 head of cattle were trailed into the Cariboo and Fraser valleys of British Columbia following the discovery of gold. When the mining market diminished the surplus cattle were driven to Idaho and then on to the San Francisco market, arriving there a year and one-half later after crossing 2,000 miles of virgin country.

But a more typical pattern of forming a trail herd, other than the large droves from Texas to the north, can be obtained from accounts in early newspapers:

"The agents of Lang & Ryan, cattle dealers, are now in Yakima and Kittitas Valleys purchasing an immense drove of cattle, which they intend to drive to St. Louis early in the spring. They expect to start with fifteen or perhaps twenty thousand head. Last year they purchased about one-fourth that number for the same market. The effect of this drain will be in a few years hence to make beef cattle extremely scarce," the Washington Standard (Olympia), January 12, 1878. And, on January 26, the "Standard" reported that Lang & Ryan bought

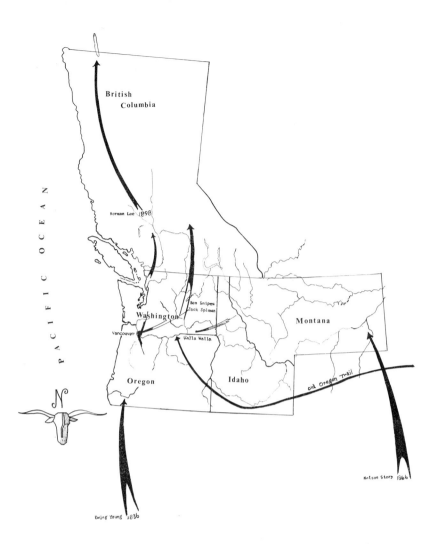

British Columbia

Noreen Lee 1898

Ben Snipes
Jack Splawn

Washington

Vancouver

Walla Walla

Montana

PACIFIC OCEAN

N

Oregon

Idaho

old Oregon Trail

Nelson Story 1866

Ewing Young 1836

Map of early cattle trails of
the Northwest.
By Virginia Paul.

12,000 head of cattle "at an average cost of $13 per head, this would give a total of $156,000 disbursed by them since the first of October."

A homesteader by the side of the trail recalled herds moving by in a continuous stream for as long as ten days to two weeks.

Unless severe hardships were encountered, the cattle could be expected to put on some weight and arrive at the destination in good condition.

There were two objectives in moving cattle — to take advantage of free grass where the cattle were fattened for market, then on to the market or, to be trailed directly to more populated areas for slaughter.

Short distances now reached by truck or rail within hours took days: "George Smith of Kittitas, the well-known cattle drover, arrived in town from across the mountains via Snoqualmie Pass on the 15 inst. having been 10 days on the road. He was accompanied by 10 white men, 7 Indians and a drove of 80 horses and cattle. He found about 10 feet of snow on the pass . . . So says the Intelligencer." published on June 20, 1880 by the Dalles Times, Oregon.

It was surprising to men of those early times, that cattle could survive the cold winters of the north. Except for an occasional winter that was devastating to cattlemen, the herds from Texas thrived in the climate — became larger and fatter. There are first-hand reports of steers 17 hands (a hand is 4 inches) in height.

Stewart H. Fowler of the Animal Science Department, Mississippi State University, provides this capsule account of the vast movement north of the Texas Longhorn:

". . . herds often numbering about 2,500 to 3,000 head were trailed from Texas through the low, hot hills of Oklahoma and on to Kansas, Missouri and the railroad for shipment to the markets of the East and (to) the ranges of such distant areas as Nebraska, the Dakotas, Montana and Canada.

"It is estimated that some ten million Longhorns were drained off the Texas ranges and driven up the north trails from 1866 to 1890 to fatten on the lush grass left by the vanishing buffalo. However, this colorful era of the Longhorn was starting to end by 1895."

60

The "Story" of Early Cattle Drives.
Nelson Story triggered the driving of cattle herds from Texas to the Northwest. Story was an ambitious man of many trades. His attempts at various occupations of the times including prospecting, logging, freighting and selling supplies, succeeded.

He realized the potential offered by the beef-hungry miners and, with this market in mind, he traveled to Texas in the summer of 1866 for a herd of Longhorns, bought 600 of them and headed north.

Though assailed by all manner of adversity—irate homesteaders, inclement weather and aroused Indians—he succeeded in reaching the Gallatin Valley of Montana before the end of the year. The achievement was a tribute to his courage and determination.

Story continued a profitable interest in the buying and raising of cattle to supply mining areas and other markets.
Courtesy Montana Historical Society, Helena, Montana.

Doan's Crossing Monument "In Honor of the Trail Drivers who freed Texas from the yoke of debt and dispair by their trails to the cattle markets of the Far North, we dedicate this stone, symbol of their courage and fortitude, at the site of the old Doan's Store. October 21-22, 1931.
The Longhorn Chisolm Trail
and
Western Trail
1876-1895"
Courtesy Texas Longhorn Breeders Association of America.

Texas Longhorns coming up out of the Red River at Doan's Crossing into Oklahoma, just south of Altusa, Oklahoma, 1966 Centennial reenactment of an early drive.
Courtesy Texas Longhorn Breeders Association of America.

There are limits to opportunity — mining markets disappeared when local cattlemen began to produce enough beef and as some ore fields were depleted; more and more land was brought under cultivation; and an idea was developed and patented by Joseph Glidden in 1874 — barbed wire.

The impact of barbed wire on cattle drives and unlimited grazing was dramatic. Direct and painful contact with the barb for man and animal was as nothing compared with the frustrations of encountering barred trails and limited access to and through the land of the Northwest.

Forage was gone along the dusty, well-trodden trails.

Trailing cattle became limited to the movement from spring to winter range, or to stockyards for shipment by rail to market.

In just a short period of time transportation progress exceeded all the improvements that occurred in thousands of years of cattle history, bringing about new marketing concepts and patterns.

The railroad came to the ranges. Within another forty years, trucks were rolling along improved road beds. Finally, markets abroad could be reached with live beef and dressed shipments by air in hours instead of weeks.

The current trend within the beef industry is to move processing plants nearer the supply and to concentrate cattle in locations better suited for feeding to finish.

Refrigeration and rigidly controlled processing have made it more economical to ship dressed carcasses rather than live animals.

The boisterous, vivid, colorful days of the great cattle drives are gone and so are the men but the lively legend that remains will never lose its universal appeal.

Jack Splawn—packer, drover, cattleman
Splawn's cattle droving experiences began when he was sixteen by helping to trail a herd from Central Washington's ranges to British Columbia's Cariboo country, a distance of nearly 800 miles. Five years later he was driving his own cattle to the Canadian Province, Idaho, Oregon and Montana. He opened new trails including one across the Cascade Mountains to the west and the Puget Sound markets.

He made his last drive from Canada to Washington State in the fall of 1896 when he was fifty-two years old. He continued to live an active, productive life and maintained his interest in the cattle business for another twenty years.

Courtesy Homer Splawn.

Texas Longhorn Trail Herd in 1880
The herd belonged to Captain Charles Schreiner. The picture was taken at Doan's Crossing on the Red River north of Vernon, Texas in Wilbarger County. Left to right; Unknown, unknown, Alex Crawford, unknown, unknown, Alex Maltsberger, Will Hale and Sibe Jones.
Courtesy Texas Longhorn Breeders Association of America.

Norman Lee was an early cattle rancher in the Chilcotin, British Columbia. When 20-years-old he left England, landing at Boston's Bar in 1882 from where he walked to Nicola.

Pooling resources with E. P. Bayliff at Cherry Creek Ranch, he headed for the Chilcotin, settling at Redstone, and two years later on the present Bayliff Ranch.

In addition, he acquired a ranch purchased from Danny Norber to which he moved when Bayliff married.

In 1898, Lee decided to try for a share of Klondike gold by turning steak on the hoof into "stakes" of gold. He started the 1200-mile drive with 200 cattle, five cowboys, a wrangler for 30 pack and saddle horses, plus a cook. Three drovers had the same idea at the same time — Jim Cornell, Jerry Gravelle, and Johnny Harris started separately with a total of 375 head of cattle.

When Lee arrived at Telegraph Creek he found Jim Cornell butchering his own beef for sale.

The next month Lee arrived at Teslin Lake where he planned to 'dress the beef out' and float the carcasses by raft to Dawson. A gale wrecked the beef laden rafts. Disaster plagued the other drives as well. The Harris and Gravelle drives "froze in" above Dawson with a complete loss of beef.

Though Lee returned without cattle or gold, he had established a reputation that enabled him to restock his ranch.

Courtesy Williams Lake Tribune.

Ben Snipes—Cattle King

From his own range in Washington Territory which bordered the Cascade Mountains on the west and extended east to and beyond the Columbia River, north to Canada and south to Oregon — Ben Snipes drove thousands of cattle to the mining markets of the Fraser River and the Cariboo in British Columbia.

During his life which ended in 1906, Snipes had known both failure and success. It is believed his cattle herd grew to 35,000 head grazing on his vast holdings. Several times severe winters claimed thousands of his cattle — a calamity shared by Northwest Cattlemen in the early days. Such losses barely dampened Snipe's enthusiasm for within days he was refinancing and eyeing prospective herds to build again. His story is paralleled by few men who started with nothing and 'drove' their way to the top.

Courtesy Dorothy Churchill

Norman Lee trailing cattle into Ashcroft in 1914 before the Pacific Great Eastern Railway came to the Cariboo.

Courtesy Williams Lake Tribune.

Dan Lee (right), son of Norman Lee, trails a herd to Williams Lake in the 1950's.

Courtesy Williams Lake Tribune.

This picture was taken in 1910, 6 miles east of Pomeroy on Highway 12. The cattle were bought on the Salmon River in Idaho and trailed to Garfield County, Washington to be fed out. There were 300 head of steers and the buyers were Billy Schneckloth, Pete Weller and Fred Miller.

Courtesy Robert R. Beale.

A cattle drive in Eastern Oregon.

Courtesy of Oregon Historical Society, Portland, Ore.

Bill and Gale Weatherly trailing cows with young calves. Asotin County, 1953.

Courtesy R.P. Weatherly.

A scene in John Day, Oregon 1955. A cattle drive from Herman Oliver Ranch enroute to summer rangeland.

Courtesy of Oregon Historical Society, Portland, Ore.

Cowboy and packhorse.

Courtesy Montana Historical Society, Helena, Montana.

J.K. "Bagger" Marsh of XIT.
Courtesy Montana Historical Society, Helena, Montana

Tom Dillard — noted cowboy and wolfer and once a Texas Ranger — photo taken 1940. "First with horses and slow with cattle."
Courtesy Montana Historical Society, Helena, Montana

Texas cowbow "Tunis" on his favorite mount in eastern Montana. L.A. Huffman photo.
Courtesy Montana Historical Society, Helena, Montana.

John H. Williams (left) when a lad of 22 came north with a string of Texas Longhorns for delivery to Nelson Story. Later he was U.S. Marshal for Montana Territory and also with '79' the John Murphy outfit.
Courtesy Montana Historical Society, Helena, Montana.

Members of a Texas Trail Herd in the year 1890. XIT Cowboys who brought some of the first XIT cattle to Montana from Texas. Upon arrival in Miles City, they went to the studio and had the picture taken. Members of crew are: back row, left to right, Steve Beebe, Frank Freeland, Billy Wilson. Front row, left to right, John Flowers, Al (Alden) Denby, Tom McHenry, Dick Mabray and Tony Mabray.
Courtesy Montana Historical Society, Helena, Montana.

N bar N driving 7,000 head 300 miles Northwest in 1892. Home, Land, and Cattle Company outfit, St. Louis, Missouri.

Courtesy Montana Historical Society, Helena, Montana

Joe Proctor, a Montana Cowboy who was born a slave in Texas. He came north in the late 1870's with one of the trail herds. It was not unusual for three or four negro cowboys to be a part of the trail crew. Proctor was a wild horse runner and a wolfer, finally settling down in eastern Montana where he was well respected.
Courtesy Evan McRae

The chuck wagon, mess wagon of the cow country, was usually an ordinary farm wagon fitted at the back and with a large box containing shelves and a lid at its rear that, when lowered, made a serviceable table. The life of a cowboy away from headquarters was always centered around the chuck wagon. It was his home, his bed and board, his hospital and office, his playground and social center. It was where he got his fresh horses. It meant fire, dry clothes, and companionship. At night it was his place of relaxation where he spun his yarns, sang his songs, smoked his cigarettes, and spent the happiest years of his life. Nothing added more to the harmony of the cowboys' life than a well appointed chuck wagon.
Courtesy Montana & Utah Historical Societies

Chuck wagon scene from the Union Pacific Collection.

Courtesy Utah Historical Society.

XIT Cowboys crossing the Yellowstone.

Courtesy Montana Historical Society. Helena, Montana

Cattle emerging from a swim across the Fraser River landing at Quesnel, British Columbia.
Courtesy Provincial Archives, Victoria, B.C.

N — (N Bar) Ranch cattle crossing the Powder River, Montana. Circa 1885. Thomas Cruse.
Courtesy Montana Historical Society, Helena, Montana

Cattle and riders swimming the Grande Ronde River in Asotin County, Washington, 1930.
Courtesy R.P. Weatherly.

Crossing the Cowiche River, Washington. An unusual photo-sequence taken in the early 1900's of cattle being ferried and used as "bait" to lead the rest of the herd across.

Courtesy Charles Gibson

"Three Sevens" Cattle (Wibaux) ready for shipment about 1895 via Northern Pacific Rail Road for Chicago Stockyards.
Courtesy Montana Historical Society, Helena, Montana

The "Horse Shoe Bar" beef herd on the way to the railroad. Picture taken September 2, 1896, on the Claggit Hill in the Bad Lands of the Missouri River.
Courtesy Montana Historical Society, Helena, Montana

Cattle being driven in U.S. Forest Reserve.

Courtesy of Montana Historical Society, Helena, Montana

Black Angus of the Deep Creek Canyon Ranch. Porter C. Fender, Belt, Montana. Photo by American Angus Association, St. Joseph, Missouri.

Courtesy Montana Historical Society, Helena, Montana

Yearling steers owned by Ervin Ellis of St. Xavier, Montana, cross the Big Horn River on their way to summer range, April 19, 1965. E.L. Frost, photo, Hardin, Montana.

Courtesy Montana Historical Society, Helena, Montana

"Flying Horseshoe" cowboys trail bovines near the mountain foothills in the eastern part of Kittitas Valley in Central Washington. This area was settled in the 1870's. Raising cattle is still the major agricultural industry. 1965.

Cattle await shipment by rail. Circa 1960.

Loading cattle in October 1967. The cattle are being moved from Hamilton, Montana to Omaha, Nebraska.

Courtesy Burlington Northern

Carnation Farms cattle being loaded to ship out of the Port of Seattle, Washington.
    History in reverse is one pattern of progress. Cattle were not native to North America. The first cattle were shipped to the continent by sailing vessels, and now, going out by modern freight ships are huge crates of breeding stock to foreign ports.

A modern trail drive in Western Montana. Though range herds are no longer trailed great distances across the land, the cattle still move many miles on the hoof between summer and winter pasture.

Courtesy Burlington Northern

Moving beef cattle from range to feedlot during a winter snow in Langell Valley, 24 miles east of Klamath Falls, Oregon.

Dale L. Swartz, photo courtesy Northwest Unit Farm Magazine

Idaho cattle drive near Emmett, Idaho.

Courtesy Northwest Unit Farm Magazine.

Moving 'em out in Oregon.

Courtesy Northwest Unit Farm Magazine.

On an Oregon trail.
Courtesy Northwest Unit Farm
Magazine

Trailing through green pastures.

CHAPTER V
Roundup, Cowboys and Chuckwagons

The work of branding began about the first of May. A dozen riders started out with a chuck wagon and a cook headed for a camp located near a corral out on the range.

The sites selected for corrals were close to good water and to favorite pasture grounds of cattle.

Out of a herd of five thousand head of cows, a rangeman expected to have around 3,000 calves to brand. If the cows received good care, the increase in calves might rise to seventy-five per cent.

Where ranges were shared by a number of cattlemen, branding was done cooperatively with men from each outfit assigned to specific duties.

A captain or "rep" (representative) was appointed from each outfit to act for the owner. It was his responsibility to see that the owner's cattle were sorted, branded and counted.

On a typical roundup day, all cattle within reasonable distance surrounding the camp would be gathered in and the cows and calves 'cutout' or separated from the rest of the herd. The cattle were held in a branding pen.

A rider would herd the calves, one at a time, toward the area where the branding was done.

A fire in which the branding irons were heated, was tended and kept burning throughout the day.

Men skilled in using the lariat, roped a calf holding it securely until one or more men had the animal branded.

A branding iron was heated to the proper degree, placed in the correct position and held on the animal for the right length of time.

Only experience could tell the cowboy when the iron was right. A grey color was the secret. The exact shade of grey meant the iron was hot enough but not too hot.

To avoid making a blotched brand, the handle of the iron was moved in a rocking motion.

The best job was done quickly when the hide was dry, the iron hot and applied only long enough to make a brand the color of well-worn saddle leather.

Several other tasks were involved in the process of branding. The animals were castrated and ear-marked. In areas where ranges were overstocked, heifers were spayed and produced good meat.

After branding all calves in the area, the outfit and wagon would move to another camp ten or twelve miles distant where branding work continued until the range in the district had been worked.

Cattle were rounded up again in the fall to separate those sold from the rest of the herd. Cattle to be sold on the ranch were counter-branded with the new owner's iron.

Such was the practice when drovers traveled through the countryside purchasing huge herds from rangemen to trail to far-away markets.

Working on a roundup was no place for an inexperienced horseman or a man disinclined toward hard work, sweat, dust or cattle. He would be closely associated with these elements for several weeks or more, depending on the size of the herd.

In the 1880's, a cowboy was paid a monthly salary of forty dollars which included twenty dollars for board. A 'Rep' or permanent, top cowhand received about sixty dollars. If extra men were hired, they were paid one dollar and fifty cents a day plus their meals. The pay was comparable in terms of the value of cattle — a year-old steer brought eight dollars — two cents a pound.

A cowboy supplied his own 'gear' which consisted of a saddle, bridle, cinch, horse blanket, bedroll and the clothes he wore.

The rancher for whom he worked, furnished him with five or six horses. The horses were unbroken (had not been ridden or trained) and it was his responsibility to train his 'string' of mounts and to see they were cared for properly. At the end of a season, the horses were sometimes sold and the cowboy supplied with 'raw horseflesh' with which to begin the training process again.

The "Padlock" roundup wagon going out for calf branding with Billy Weeks, the cook, driving the chuck wagon in the lead, and Verne Torrence driving the bed wagon. "Cavvy" or remuda of saddle horses follow. Owners are the Scott Land & Livestock Company of Dayton, Wyoming. Range on Crow Indian Reservation, Montana. July 1965. E.L. Frost photograph.

Courtesy Montana Historical Society.

A cowboy was known as the man with the 'guts and a hoss.' He rode and roped his way to glamor in early Western history.

He worked long hours, spent weeks in the saddle and his accumulated wages on a weekend in the nearest town.

He was generous with his companions and loyal to his 'boss' and branding iron. Quick action was a part of his life and came before thoughts of his own safety.

Men assigned to go along with shipments of cattle by rail were called 'cowpokes.' Their task was to check cattle cars when the train stopped and to prod any cattle that were down to a standing position. This helped curb losses during cattle shipment.

To call a cowboy a "cowpoke" was considered an insult.

Cowboys were proud of their attire and equipment, both of which showed a Spanish or Mexican influence in design and adornment.

It was said a cattleman or 'boss' could be easily identified because his saddle was old, used and abused and his clothing was worse than a cowboy's castoffs.'

Western attire is no longer limited to the range or ranch, but experience is as necessary as it was in the past.

When the last herds of the large outfits in Montana, and throughout the Northwest, were dispersed as free range disappeared, cowboy crews were pared

accordingly. Many exchanged life on the range for the rodeo circuit. What had once been a favorite pastime, that of testing cowboy skill to see who excelled, became a way of making a living. Such was the case of Casey Barthelmess, a Miles City, Montana, cowboy.

Casey challenged the wildest broncs to unseat him. "In those days, wild horses were mounted in the arena, not in a chute," Casey said (A feat no longer required.) Casey achieved well-earned recognition for his exploits in the early day rodeo arena.

Charles Russell captured the Western range and ranch scenes and the life of the cowboy on canvas in a vivid, exciting manner that will never be excelled. With pen and brush, he portrayed the colorful era the way he saw it and lived it, with zest and abandon.

Today's cowboys 'head and heel" with the skill of one hundred years ago and clouds of dust hover over the branding corrals in the fall. Branding fires burn and irons are heated grey-hot. Smoke and the smell of singed hair still sting the nostrils. The sound of bawling cattle echoes in nearby hills.

The ranch pickup truck brings a hot, noon meal prepared on the kitchen range. There are exceptions where the old chuck wagon and cook seem best.

As long as cowmen and cowboys are inclined toward hard work, dust, sweat and cattle, there will be a roundup at branding time.

On the roundup — Saddling up.

Courtesy Montana Historical Society

Rounding 'em up, vividly portrayed in this long ago scene. Cattle were gathered as they grazed the range. Cattle often belonged to more than one owner, who were represented at the roundup to tally the number of head and ensure the proper brand was applied.

Courtesy Montana Historical Society

Roundup time in the hills south of Kamloops, B.C. around 1900. The cattle would likely be Lew Campbell's. Bill Roper's, Johnny Hull's, and McLeod's, all old-time cattlemen in the district.

Courtesy Provincial Archives, Victoria, B.C.

Handling a bunch at the head of a creek. L.A. Huffman photo.

Courtesy Montana Historical Society.

Near Roseburg, Douglas County, Oregon.

Courtesy Oregon Historical Society

With the roundup — Judith Basin 1888.

Courtesy Montana Historical Society

Roundup in Judith Basin. C.M. Russell in picture.

Courtesy Montana Historical Society

Close view of Roundup. L.A. Huffman photo.

Courtesy Montana Historical Society

Near Klamath Falls, Oregon.

Fall roundup, Boise Valley, Idaho. Early 1900's.

Cutting out a steer. L.A. Huffman photograph.

A remuda on a ranch in Montana. Circa 1921.

A 'cavvy' of 185 head of saddle ponies in the rope corral of the XIT outfit at 17 Mile Creek, North of Glendive, Montana. The photographer says he has never seen a prettier bunch of horses on the range.

Courtesy Montana Historical Society

Cowboys catch their mounts for the afternoon's work at branding time on Padlock range, Crow Indian Reservation, Big Horn County, Montana. The "Padlock" is operated by the Scott Land & Livestock of Dayton, Wyoming. Left to right are Charley Secrest, Barry Roberts, foreman, Don Redding, June Redding, Walt Secrest, Bob Fitzpatrick, (partly hidden) and Lee Secrest leading the pinto. June, 1951. E.L. Frost photo, Hardin, Montana.

Courtesy Montana Historical Society

The "Milliron" Saddle Stock held in a rope corral while the cowboys have their midday meal (usually about 10:00 a.m.) at branding time. The cook-tent and chuck wagon (under flu) in background and bed-tent and bed wagon in middle ground. The "Milliron" is owned and operated by Harvey Willcutt of Hardin, Montana. The range is on the Crow Indian Reservation, Big Horn County, Montana. July 1964. E.L. Frost photograph.

Courtesy Montana Historical Society

Removing ovaries (spaying) from a cow. Note the man with hand inside the cow. Picture was taken in 1910 on Blackfoot Indian Reservation. The man on horse to right is Joe McKay, a half breed Indian. "I rode on the bed wagon with him for two seasons. He was the night hawk in charge of saddle ponies." G.V. Barker photo.

Courtesy Montana Historical Society

E.C. (Teddy Blue) Abbott and
Chas. M. Russell.
*Courtesy Montana Historical
Society*

"TN" Outfit on Tongue River, U.S. Military Reservation. 1. Taylor 2. Frank Bircher 3. George
Vernier 5. Patty Ryan 6. Deloss McBride 7. Taylor.
Courtesy Montana Historical Society

Shoeing a bronc.

Courtesy Montana Historical Society

Joseph Creek cattlemen, 1930. Ben Tippett, Pete Edgmand, Joe Bly, Jidge Tippett and Clarence Spangler.

Courtesy R.P. Weatherly

Cowboys on the roundup. Miles City area.

Courtesy Montana Historical Society

Circle Roundup Herd. Left — Claude Allan, Right — Sam Kelly. S.W. of Glasgow, 1906.
Courtesy Montana Historical Society

With the Circle outfit in Valley Company. Harry Martin on "Crappy," Jim Ogle on 'Old K'.
Courtesy Montana Historical Society

Early day Rough Riders of Okanogan County around 1898. Left to right are 'Indian' Edwards, 'Birdie' Allison, Hans Richter, Tom Caphola, Peter Swimpkin, Heneas Charlie, Witchal "Dan" Allison.

Courtesy Ross Woodard

Left to right: Percy Robinson, Oscar Daughtery. Cowboys: "N-Bar" Ranch — Musselshell Valley, Thomas Cruse owner. 1903 Brand book shows ranges in Fergus & Dawson counties.

Courtesy Montana Historical Society

"79" top brass in 1909. Extreme Western Camp near foothills of the Crazy Mountains. Left to right: "Wild Bill" Sutter, Walt McCool (Wagon Boss), Rollie Heren, Manager. John T. Murphy outfit.

Courtesy Montana Historical Society

O.D. Gibson Cowboys. Left to right: Charles, Lymon, Ellsworth and O.D. Gibson, father, in 1912. Washington.
Courtesy Charles Gibson

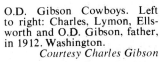

John W. Snook and sons, Fred and John, mounted for round-up. 1922. Idaho.
Courtesy John W. Snook

The Cotton Ranch Indian cowhands, probably Chilcotin. "They did everything." British Columbia.
Courtesy R.A. Moon, Williams Lake, B.C.

Rolling Satus hills, Klickitat County, Washington, have provided range for cattle herds since the middle 1800's. W.L. Coleman and Lowell Shattuck covered many miles together on horseback. Circa 1924. Wash.
Courtesy W.A. Coleman

An Indian crew on 150 Mile Ranch. Left to right: Jimmy Sandy, Tommy Wycotte, Andrew Gilbert, Tommy Wycotte, Jr. and Robert Gilbert. 1928 B.C.

Courtesy Mr. & Mrs. Hugh Cornwall

Cowboys on 150 Mile Ranch. Left to right: Unknown, Orville Fletcher, Clarence Zirnhelt and Spencer Patenaude. 1928. British Columbia.

Courtesy Mr. & Mrs. Hugh Cornwall

Dick Ranch cowboys, 1935. Sonny Polumsky, Curt Brooks, unknown, Len Brooks and Ralph Brooks. Washington.

Courtesy R.P. Weatherly

X Roundup outfit near Fallon, Montana in 1886.

Courtesy Montana Historical Society

"N Bar" Ranch mess wagon. Spring roundup. Circa 1898.

Courtesy Montana Historical Society

Cowboys at mess, Dawson County, 1898.

Courtesy Montana Historical Society

Water Tank Roundup — Circa 1910.

Courtesy of the Montana Historical Society.

Harney Company camp. Gifford photograph.

Courtesy Oregon Historical Society

Conrad Circle Cattle Company. Roundup outfit in Valley Company 1906 Circle Roundup Outfit No. 4. "Enough men eat the beef raw where generally it was a two-man job."

Left to right standing: Jimmie Ogle with knife "75" cowboy; Tommy Smith flunky; Bob Aters "J" cowboy; Charlie Purcell and "Post and Rail" cowboy; Claude Allen cowboy — Wigmore Cattle Company; cowboy J.C. Wigmore, Manager.

From Left — skinning beef: Tom Manis cowboy: Bob Malone cowboy; Matt Morgan — Circle cowboy; Harry Martin cowboy.
Courtesy Montana Historical Society

Noon camp of the "Circle Diamond" roundup wagon Circa 1900. Bloom Cattle Company, Trinidad, Colorado. Foreman, John Survant, Malta, Montana.
Courtesy Montana Historical Society

Cattle and Cowboys "IX."

Courtesy Montana Historical Society

Roundup with men around the chuck wagon.

Courtesy Utah State Historical Society

Cowboy camp, North Fork of Milk River, 1894.

Courtesy Montana Historical Society

Judith Roundup — 1885-1886. Utica, Montana in background. Back Row: 1. Jack Murphy — standing. 3. Dan Martin — seated. 4. Alec Tuttle. 5, 6, 7, 8, 9. Unknown. 10. Jack Flynn, "Cook" — standing. 11. Tom Waddell — standing. 13. Gene Gray — standing. 14. Joe King — standing. 15. Henry Gray — standing. 17. L.B. Taylor. 18. Henry Kauffman — last in backrow. Front Row: 1. Zack Whitcumb. 3. Charles M. Russell — artist. 4. Pres Larcum. 7. Johnny Sellars. 8. Henry Gates. 9, 10, 11, 12. Unknown. 13. Terry McDonnell. 14. Jesse I. Phelps. 16. Charlie Mattson — on end.

Courtesy Montana Historical Society

"XIT" Cowboys.

Courtesy Montana Historical Society

Chuck Wagon and "H S" Cowboys, 1886.

Courtesy Montana Historical Society

Barber Shop on "Circle" Roundup Camp, south of Glasgow
in 1906.
*Courtesy Montana Historical
Society*

Evening at the roundup at Big Pumpkin Creek. L.A. Huffman photo.
Courtesy Montana Historical Society

The roundup breaking camp taken by L.A. Huffman, an early photographer who captured and recorded the tempo and times of the boisterous West in Montana at the turn of the century.

Courtesy Montana Historical Society

"The Last Roundup on the Range," Last trip of the "XIT" Cattle Company — "Gathering and shipping all their cattle and going out of business. This was in 1908 at the forks of Burns Creek west of the Bad Lands and south of Yellowstone Valley — 20 miles north of Glendive, Montana. A cavvy of 165 thoroughbred saddle ponies, the finest bunch on the range. Sixteen men all in sight. Was forced to take this against the sun with white dust screen in back. Which shows more horses than usual. Rufus Morse was man in charge in lead beside cook wagon." G.V. Barker photo.

Courtesy Montana Historical Society

Early fall cattle roundup on the Beckley and Killingsworth Ranch near Coulee City, Douglas County, Washington. The calves will be separated from the cows and sold. Ernest Busek photo. 1966.

CHAPTER VI
Free Range to Fences

This is a big land stretching from beyond the Rockies to the Pacific Ocean and from Alaska to the California border. A land of bunchgrass. forest-covered mountains topped by jagged, rocky crags, a land of pine and sagebrush, swales and meadows. It was, and is, a land of individualists.

In this vast rangeland, unbroken by fences, wild horses once ran free. Where buffalo were abundant, cattle grazed the native grasses.

The 30-year period beginning in the 1860's was a time of independent, stubborn men, growing cattle herds and turbulence.

Rangemen for the most part, shared a respect for each other and their property, a dislike for sheep which they claimed ruined the grass, and a deep resentment of homesteaders. Tempers flared, sometimes erupting into bloody violence between individuals and groups.

This was a time when a single cattleman's range included 12,500 square miles or more. Land was held by pre-emption, homestead entries and by first occupation. There was very little land purchased except for agricultural purposes. Stockmen shared ranges because cattle were so numerous it was impossible to keep them separated.

Cattle drifted from summer to winter ranges reaching the winter ranges around the first of November and wandering to spring ranges early in May. Little hay was harvested for winter feeding. Cattlemen considered it impractical to cut and store hay because in times of deep snow, it was impossible to get the hay to the cattle and because the cost of 'putting up' hay did not seem justified.

The natural grasses and forage included bunchgrass, rye grass, white sage, cane grass, alfileria, willow, ross, greasewood and rushes in wet land, among other native plants.

In many range areas water was extremely scarce.

In the latter part of the century rangemen realized the days of the big herds were over. The country was overstocked. Additional pressure was being applied on ranges as arable lands were homesteaded and cultivated. Stockmen were bringing sheep to the ranges and cattlemen claimed the ranges were being permanently injured by the "pestilence." "There should be a United States law confining them to certain limits, from which they should not encroach upon cattle ranges and take the life out of the enterprise," reported Clarence Gordon in a 'Supplementary to Enumeration of Live Stock on Farms' in 1880.

Laws having an impact on range use included the 1862 Homestead Act which abetted settlement by offering 160 acres of land in return for residence and cultivation. What was not realized at the time was that marginal land could not provide sufficient returns for continued operation of the property as a means of livelihood.

An Act of Congress in June of 1897 legalized government authority over grazing. In the beginning, the land policy of the government had been more concerned with land disposal through the Homestead Act, land grants to railroads, schools and institutions. The principle behind the policy was made with good intentions, to encourage settlement, improve and develop conservation and further the development of transportation and education.

Lands previously claimed by individuals were made available for grazing only to those qualifying for a permit. Prior use by the individual, need for the range and the amount of property owned privately, were taken into consideration. No charge was made for grazing privileges and for all intents, it remained "free grass."

Also in March of 1891 more land was removed from grazing when the President of the United States was empowered by Congress to create Timberland Reserves setting aside part of the public domain.

Pete French came to Harney County, Oregon, from California at the request of Dr. Glenn, with whom he was associated and who had visions of establishing a cattle kingdom in the state to the north. French's and his partner's hopes included finding a calf pasture 'so big it would run for 60 miles.' To implement the idea, French had his ranch employees make homestead entries that were adjoining, one-quarter mile wide and one mile in length. The land chosen lay in the Blitzen Valley extending to the western side of Steens Mountain.

One side remained open and it was in this section that a settler filed a claim. This touched off a controversy between big and small ranchers and homesteaders that ended in the death of French at the hands of a settler named Ed Oliver.

Accounts vary as to the exact date and events surrounding the incident but it is a fact that Pete French died of a gunshot wound in the late 1800's ending the career of an ambitious, successful livestock organizer. He had gathered together and held a cattle empire, supervising an immense domain totaling some 160,000 acres.

The land later became part of the Malheur Wildlife Refuge reverting to its natural, wild state, much as it was before the time of Pete French, cattleman.

Courtesy Oregon Historical Society, Portland, Oregon

Cattle in the Quesnel District, British Columbia.
Courtesy Provincial Archives, Victoria, B.C.

In 1906, a seasonal fee was placed on public-owned grazing land based on the number of animal head. This was changed to a per animal unit month (a unit is based on the amount of land to feed one animal one month) in 1927 and later in 1933, it was adapted to reflect cattle prices.

Throughout these changes it was the cattlemen who used the land for grazing and who became the major partner with the government in the more liveable areas. In 1916 a stock raising provision permitted 640 acres, or a section one mile square, and eliminated the crop-growing section, for without a ranching enterprise on marginal lands, success was not likely.

The Taylor Grazing Law was enacted in 1934 banning settlement of Public Land prior to classification by the Interior Department. The act was designed to make better use of public lands, stabilize the livestock industry and facilitate the establishment of districts, permits and fees.

Twelve years later the Bureau of Land Management was created under the Department of Interior which consolidated the Grazing Service and the General Land office which had been established in 1812.

In Canada, the Land Act of 1872 provided that anyone 21 years of age or older could file for a quarter section on even-numbered sections for a fee of $10.00, three years of occupancy and improvement. Crown land could be purchased in addition.

Nine years later a regulation in the Act allowed for 21-year leases up to 100,000 acres per lease at one cent for each acre. The lessees were required to stock the ranches with one head of cattle per ten acres within three years. The impracticability of this regulation became apparent and in a short time the regulation was relaxed to require one head per twenty acres. One of the inducements offered was duty-free importation of cattle from the United States by the lessee.

Throughout the Northwest, ranchers working cooperatively with State, Federal and Provincial Grazing and Forest Departments, continued to improve rangeland by seeding, maintaining fences, creating water reservoirs and controlling brush growth among other approved techniques to increase the potential and to better conserve natural resources.

Productivity of private lands has been increased, easing the burden of public lands.

In more recent years there has been great public pressure for more recreation areas and for a reduction in the amount of public grazing land.

Since tomorrow's pattern of living is affected by today's decisions, the future use of land should be considered carefully. There is a direct relationship between land and food. To ensure a plentiful supply, enough land must remain for agricultural use.

Con Shea's watering hole two miles above the Guffey railroad bridge near Bruneau on the Snake River. (A very old site.)

Courtesy Idaho Historical Society, Boise, Idaho

Range scene in Steens Mountains, Harney County, Oregon.

Courtesy Oregon Historical Society

Wallis Huidekoper's cattle, Eastern Montana and Dakota.

Courtesy Montana Historical Society, Helena, Montana

Part of the "Circle Diamond" herd on water at Freewater Crossing on the Milk River about 1902. Bloom Cattle Company.

Courtesy Montana Historical Society, Helena, Montana

An early picture of Hereford cattle on the open range near Fairfield, Idaho.

Courtesy Idaho Historical Society, Boise, Idaho

Biering Cunningham horses on a lake in Taylor's Fork of the Gallatin River, Montana in 1910.

Courtesy Montana Historical Society, Helena, Montana

Biering Cunningham cattle on Cash Creek Basin, Gallatin Madison Divide, Montana, 1910.
Courtesy Montana Historical Society, Helena, Montana

Cattle on the range on Camas Prairie around 1910.
Courtesy Idaho Historical Society, Boise, Idaho

114

Burnt River Cattle Ranch near Baker City, Oregon.
Courtesy Oregon Historical Society, Portland, Oregon

Cattle grazing. U.S. Forest Service Photograph.
Courtesy Montana Historical Society, Helena, Montana

Two-year olds on Toats Creek Road, 1920. Okanogan County, Washington.
Courtesy Ross Wooward

On the Sinlahekin Ranch in 1920.
Courtesy of Ross Woodard

A ranching scene in the Kamloops District, British Columbia.
Courtesy Provincial Archives, Victoria, B.C.

Scab rock range is good spring grazing. It is apparent the cattle are not used to photographers. Cattle are uniform but not as blocky as later herds became. 1922.

Courtesy Northwest Unit Farm Magazines

Cattle on the Simcoe Range of Satus in 1927. Washington.

Courtesy W.A. Coleman

Jim Dexter and Ross Woodard saddling up in front of the original Guy Waring cabins, Sinlahekin Ranch

Courtesy Ross Woodard

Trailing cattle to the summer range at the head of Yankee Fork Drainage (McKarys Creek). August 1930.

Courtesy Lawrence F. Bradbury

Cattle in Eastern Oregon. Arthur M. Prentiss photograph.

Courtesy Oregon Historical Society, Portland, Oregon

At the base of Steens Mountain.

Courtesy Oregon Historical Society, Portland, Oregon

Cattle.

Making the last ride before winter closes in. Ross Woodard, Dick Hill, Johnny Johnson and Jim Dexter at the mountain cabin in November 1930.

Prescott Ranch, Chester, Liberty County, Montana.
Courtesy Montana Historical Society, Helena, Montana

Don Hunter, Big Blackfoot Valley, November, 1943.
Courtesy Montana Historical Society, Helena, Montana

Cattle on pasture. Whidbey Island, Washington. This pasture was once farmed, producing about 8 bushels of oats to the acre. Ample rainfall in this area produces lush grass. 1947.

Courtesy Northwest Unit Farm Magazines

Cattle on the home ranch of Lawrence F. Bradbury, Challis, Idaho in 1957. The ranch was started by Bradbury's grandparents in the late 1870's.

Courtesy Lawrence F. Bradbury

The W.J. Dorman herd of polled Herefords grazing on a fine stand of native bunch grass near Helix, Oregon. 1948.

Courtesy Northwest Unit Farm Magazines

Bringing 'em in from the meadow on 150 Mile Ranch British Columbia in 1948.

Courtesy Mr. and Mrs. Hugh Cornwall

Broad view of the BX rangeland at Vernon, B.C.
Courtesy Provincial Archives, Victoria, B.C.

Spring in Jordan Valley, Oregon. 1947.
Courtesy Northwest Unit Farm Magazines

Line Riders

Joe Wheeler (left) and Oscar Kamhost, line riders pictured at the Ranch on the Yellowstone in February. 1891. (The brand belonged to the Northern Cattle Company with ranges on the Powder River, Little Powder River and tributaries and the Tongue River and tributaries.

A line rider is a cowboy assigned to keep fences between boundaries in repair. Line fences seem to have no beginning or end, stretching for miles down through gullies and up over hills and ridges — a thin thread pointing to the horizon and then continuing on into the distance.

Line camps were established on the range usually within a day's ride. The camps consisted of a small log cabin or frame shack, a stove and a bed. There was a small corral and a lean-to for saddle horses. There were times when a man's saddle was his pillow and a blanket his shelter.

Line riding was lonely. Human companionship was limited to other line riders or to coming across a sheepherder's camp when loneliness might overcome a cowman's natural reluctance to have anything to do with anyone connected with sheep.

Where timber was available, early fences were of wood, split rails or poles. After barbed wire was introduced in the West, it was strung along slim poles thrust in the earth if there was enough soil or supported by slanted poles if there was not.

Any person who cut a line fence or left a gate open was considered of mighty questionable character. Line fences are still maintained and to the casual observer lead to nowhere. And those who damage them are held in no higher regard than in days gone past.

Courtesy Montana Historical Society, Helena, Montana

Rail Fences.

Courtesy Northwest Unit Farm Magazines

Pole fence on land in the Quesnel District at Dragon Lake.

Courtesy Provincial Archives, Victoria, B.C.

Cattle country near New Meadows, Idaho. Bureau of Reclamation Photo.
Courtesy Bureau of Reclamation Department

Robert L. Ross and wire col-
lection.
*Courtesy Northwest Unit
Farm Magazines*

The pastures are divided by an electric fence for rotation grazing. The cattle are shown in one of the pastures starting a grazing period.

Courtesy U.S. Department of Agriculture, Soil Conservation Service

Cattle, North Side Pumping Division, Minidoka Project, Idaho. Leonard Kraemer raised a fine herd of Black Angus Cattle by using the strip pasture technique. This practice involves the dividing of the pasture with single wire electric fence and moving the cattle from one section to another before the pasture gets eaten too low. The pasture then grows back faster and more tonnage is realized per acre.

Courtesy Northwest Unit Farm Magazines

C.R. Lampheir and gate closer.

Courtesy Northwest Unit Farm Magazines

127

DOUBLE BRACE
MAXIMUM STRETCH DISTANCE 80 RD.
POSTS SET 3.5 FEET
THIS TYPE BRACE STRONGEST

SUSPENSION FENCE
POSTS 80 TO 120 FEET APART
MINIMUM 9 INCH SAG
NEED EXTRA STRONG BRACE
ABOUT ½ COST OF STANDARD FENCE

Ray Morris, WSU County Extension Agent assigned to the Colville Indian Reservation, kneels by the double brace system required in a good suspension fence. He uses suspension fences around his grass fertilization demonstration plots. Suspension fences cost about one-half as much to build and maintain as a standard fence. *WSU Photo.*

Cattle on the "Walking T" Ranch then owned by Alan Rogers, Central Washington Cattleman, recognized for his ranch development and contribution to the welfare of the cattle industry. 1949.

The Reinbold Ranch range near Davenport, Washington, 1948.
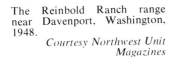
Courtesy Northwest Unit Magazines

The first settlements in the Northwest were along rivers and streams. The most valuable land is still that land bordering natural sources of water, for practical as well as esthetic reasons. This cattle scene was taken along the Little Spokane River. Washington in 1949.

Courtesy Northwest Unit Farm Magazines

129

Eyeing the beef herd grazing on the grassy, rolling land of the Douglas Lake Ranch, British Columbia.

Courtesy Provincial Archives, Victoria, B.C.

The Harder range a few miles south of Sprague, Washington. June 24, 1951. Photo by C. Hagen.
Courtesy Northwest Unit Farm Magazine

Rangeland near Sprague, Washington. Photo taken by Hagen in 1950.
Courtesy Northwest Unit Farm Magazines

Grassy range on the A.E. Reiman Farm at Banks, Idaho. 1952.
Courtesy Northwest Unit Farm Magazines

Cattle on irrigated pasture along Wild Horse Creek near Adams, Umatilla County, Oregon, the Herefords owned by Orval and Marshall McCormmack. All grazing land seeded to Brome, Alta, Orchard grass, Tall oats, and Ladino Clover and carrying 60 steers on 30 acres. 1951.
Courtesy Northwest Unit Farm Magazines

Making better land and beef are among the challenges Silas Brainard faced on his ranch in the Chimacum area, Lewis County, Washington. 1953.
Courtesy Northwest Unit Farm Magazines

Bunchgrass is native to the prairie country of the Northwest. Overgrazing in the late 1800's and early 1900's destroyed some of the natural grasslands, which allowed an increase of sagebrush. Since bunchgrass contains nutritive qualities and is well adapted to the climate, efforts are being made to increase production. The grass on this range near Lacrosse, Washington measured 20 inches in the spring. Mr. L.S. Branch shown measuring the stand. Land was purchased by Hinderer Brothers.

Clarence Kelly photo.

Ross Woodard's favorite picture taken in 1960 of cattle trailing to summer range from the spring range.

On a ranch near Cheney, Washington, Robert Grogan (right) works his herd of Hereford cattle on Bluebunch wheatgrass and Idaho Fescue. 1967.

Clarence A. Kelly photo.

A "watering hole" in forest land. Most federal and state lands are operated under a multiple-use plan. This allows for cattle-grazing without depriving native ruminants — deer and elk, of feed. No doubt, the pond is shared, also. U. S. Forest Service photo.

Courtesy Northwest Unit Farm Magazines.

A most desirable source of water for pasture or range is a natural stream. Montana cattle gather at the edge.

Courtesy Burlington Northern

Following a sagebrush burn ten miles north of Edwall, Washington on the Scott Barr Ranch, Barr (right) and Len Dupier examine the airplane seeding of Intermediate Wheatgrass, Crested Wheatgrass, Hard Fescue and Big Bluegrass.

Courtesy Northwest Unit Farm Magazines

Herefords cool off in Oregon's cattle country on a hot July day in 1965. Cattle thrive better with protection from the elements of summer and winter.

This 6000 gallon tank stores water from where it is pumped at McCarter springs, 25 miles northeast of Madras, Oregon. From the tank site, water is supplied to three troughs on a gravity basis. The striped pole seen on the tank is a gauge which indicates the depth of water in the tank. The gauge permits checking the amount of water in the tank by reading it with binoculars. Vandervelden photograph. Soil Conservation photo 1968.

Courtesy Northwest Unit Farm Magazines

The development of a stock pond permitted cattle to range upland pasture until late September. There were no other water facilities on this range in Kittitas County, Washington, before development.

Courtesy Northwest Unit Farm Magazines

Francis Ripplinger and horse drinking from a new spring development in the Teton watershed 4 miles west of Driggs, Idaho, three thousand feet of plastic pipe were installed. Richard L. Thompson photo.

Courtesy Northwest Unit Farm Magazines.

This water tank is another step in bringing water to the animals, guarantees a constant supply.
Courtesy Northwest Unit Farm Magazine

White-faced steers grazing on a hillside ranch near Salem, Oregon, create a peaceful, pastoral scene.

Stock pond on the Webb Ranch, Camas Prairie

Courtesy Northwest Unit Farm Magazine.

Cattle feeding from Orchard-grass pasture on fine, sandy loam near Moses Lake, Washington on land operated by Bob Hardin.
Courtesy Northwest Unit Farm Magazines

Cattle grazing in the Sawtooth Basin north of Sun Valley, Idaho.
Courtesy Northwest Farm Unit Magazines

The country around Salem, Oregon is well suited to the production of beef.
Oregon Department of Agriculture photo.

Black Angus pasturing tall Wheatgrass in Whitman County. Eastern Washington. Roe D. Crabtree photo.
Courtesy Northwest Unit Farm Magazines

Guarded by a stone bluff. Oliver Ruens herd grazes on ample pasture near Clarks Fork, Idaho.
Courtesy Northwest Unit Farm Magazines

Black Angus cattle grazing on one to two-year-old seeding of tall Wheatgrass at Hotcreek, near Twin Falls, Idaho. Frank Roadman photo.
Courtesy Northwest Unit Farm Magazines

CHAPTER VII
Mark of Man

It is a proud moment when a man makes his mark on the hide of a beef animal. That mark declares ownership, labels a man's integrity, and tags him with responsibility. One after another, it totals his worth.

A hot iron, made in various shapes, has been evidence of possession for centuries — since the time of the early Romans and Egyptians. The first brand in America was stamped on Spanish cattle and horses brought to the continent by Cortez, Three Christian Crosses.

A brand can be a number, an initial, a letter, a symbol, a slash, a dot, a bar, a circle or a part of a circle, a diamond, a box or a triangle. It can be leaning, lazy, walking, crazy, broken, flying or reversed, hanging or connected.

It can be any one of these or a combination and is read from left to right or from top to bottom.

A brand reflects the humor and ingenuity of cattlemen and its construction, the skill of the blacksmith, cattleman or cowboy who devised it to leave a precise, legible imprint on the hair and hide of an animal.

The major purposes of branding are to declare ownership and to prevent theft, a problem that has always plagued cattlemen.

Brand recording began in the Northwest in the early 1850's. Prior to that time, a cattleman's brand was HIS because he originated it and used it.

Throughout the domain, brands were used as a means of establishing herd ownership by ranchers who shared rangelands. It was possible to increase a herd by branding a wandering calf. In the words of an early cattleman: "The number branded depends wholly upon one's vigilance. Some men will brand more calves than their cows number; others, of course, do not get all their increase branded."

The situation continued into the 1880's as another cowman testifies: "Many calves are necessarily missed, and when these leave their mothers, or are weaned naturally, they are called "slick-ears," "sleepers" or "mavericks," and belong to any cattleman who can get his brand on them. In the spring men go out "Slickearing" with lassos and branding-irons on their saddles and secure such calves. The animals are roped and the iron put on, having been heated over a fire of sagebrush or cow-chips. This sort of business has proved most profitable to some cattle-raisers, as they brand more calves than their cows number."

Sometimes a running iron was used out on the range for branding, legitimately or otherwise.

The running iron was a circle of iron which was heated, then held between the prongs of a forked stick or two sticks in applying the brand. The iron was turned, using part of the circle to create the brand design.

There were men who became experts in using a running iron to make the brand and in reworking a brand on an animal into a different brand.

When brands were changed for devious reasons, a cinch ring from saddle gear was sometimes used to avoid being caught with incriminating evidence.

To help curb such practices, a system of brand recordings was established, but surveillance continued to be the best retardative.

Many of the early brand records have been destroyed and forgotten because territorial, district and county boundaries were reshaped by political and legislative process as the Northwest grew into statehoods.

In 1849 the Oregon Territory included the present states of Washington, Oregon and Idaho, parts of Montana, Wyoming and British Columbia.

Washington was created a separate territory in 1853 with boundaries designated as all north of the Columbia, of the 46th to the 49th parallel and from the Pacific Ocean to the Rocky Mountains.

An early brand in the North West. The brand of the Lewis & Clark original expedition. Iron was used to make this copy which is now in the collection of the Oregon Historical Society.

Courtesy of the Northwest Unit Farm Magazines

Unique brands used between 1864-69 in Washington Territory.

Photo by author.

Walla Walla County included "all of the territory east of a line north from opposite the mouth of the Deschutes River to the Canadian Border, west of the summit of the Rocky Mountains and north of the Oregon Territory."

Idaho Territory was established in 1863 and Montana the next year. British Columbia joined the Canadian Confederation in 1871.

One of the earliest brands to be recorded was in 1852 at Olympia, Oregon Territory.

Having a brand recorded did not necessarily mean the recorder was the only rancher in the particular area using the brand.

The British Columbia Brand Book alludes to the duplication: "Cattle and Horse Brands have not always been centrally registered in Victoria; originally they were registered in the District Magistrate and later on the Government Agent. The exact same brand could have been registered to more than one owner in different districts."

As early as 1854, the Oregon Territorial legislature enacted a brand-recording law requiring the county clerk of each county to record, on application of any person, a description of brands and marks of livestock.

Though recordings were made in the 1850's, brands were used in the territory prior to that time and many originated years before.

"The Swaggarts of Eastern Oregon employ a branding iron which has been used since 1812, that beats anything in old brands this corner has heard about. The facts about this 135-year old implement came to light after this page published an article about the remarkable collection of branding-iron imprints gathered and used as "mural" material by the Krouse machine shop of Pomeroy. Frank Swaggart of Athena immediately mailed a description and history of his family brand to the Krouses after reading the article.

The brand is a numeral "2" over a bar. Of it, Frank Swaggart told the Krouses:

"Nelson Swaggart used this brand in Kentucky in 1812; then in Illinois in 1846; then to Eugene, Oregon, and from there to Weston, where it was used on the Swaggart homestead in Lamar gulch. It was used by Link Swaggart, and Milt and Ben. I have used it 50 years." Walla Walla Union Bulletin, May 5, 1947.

The "Balloon Bar" brand, another old brand that has seen continuous use since the early 1800's, has an intriguing background.

It was brought to the Willamette Valley in Oregon Territory by Mrs. Philip Painter in 1850.

She and her husband and family of seven children left St. Genevieve County, Missouri, under military escort to join her father, Major Moore, who was already in the Northwest.

The cattle and horses they brought with them were branded with the "Balloon Bar" Brand.

On the trail, Philip Painter and two sons died of cholera leaving the mother, two daughters and three sons to continue on to the Pacific Coast.

Display of Joseph Creek branding irons, 1952.
Courtesy R.P. Weatherly

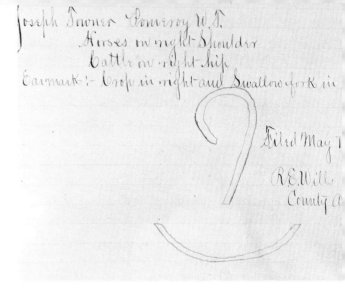

Copy of a brand recording in the old Garfield County brand book by Bob Beale of Pomeroy. The record of these brands started in Garfield County, Washington in 1881 very shortly after the county was formed.
Courtesy Robert R. Beale

One of the daughters, Margretta, married David Schnebly. The Schneblys used the brand in Walla Walla County where they established a cattle ranch, moving to the Kittitas Valley (Washington Territory) in 1872. The brand is in current use and is recorded in the family name.

When brand recording changed from a county and district level to a central location, not all cattlemen were able to retain ownership regardless of the age of the brand and amount of use. The first rancher to record the brand became the owner for as long as he chose to record it. This caused some problems.

Tom Drumheller, Ephrata, Washington, recalled the indignation of his father and uncle on learning the "Railroad Brand" had been recorded previously by another party. Their concern over losing the brand, which they purchased from Lord Blythe, was due to the fact they had just branded 3500 head of cattle not knowing the slanted, long double slashes were no longer theirs.

Mr. Wigenshaw of Okanogan County was similarly distressed by losing the "Triangle" which he had used for sixty years. In making his application, he discovered a Mr. O'Neil had already recorded the brand. Since two identical brands could not be recorded in Washington State, it was recommended that a Dot be placed in the Triangle, a suggestion that satisfied Wigenshaw.

British Columbia established a "Brand Act" called the "Cattle Ordinance" in 1869: "An Ordinance for the better protection of Cattle, and the better prevention of Cattle Stealing," dated March 9, 1869.

The word "Cattle" was defined as extending to and including "Horses, Mares, Fillies, Foals, Geldings, Colts, Bulls, Bullocks, Cows, Heifers, Steers, Calves, Sheep and Asses."

The "D" brand is said to be the oldest brand in Oregon and was owned by Samuel Dement of Myrtle Point in Southwestern Oregon. In 1972 the brand was still recorded to the Dement family. Dement came to Oregon country in the gold rush of 1848, first using the brand on five elk he had gentled and used as beasts of burden, then traded for beef cattle.

Other old Oregon brands include the "Wrench" or "CSC" brand of Miller and Lux, which the partners obtained when they acquired a California stage company's holdings.

Oregon's state brand recorder reported "none of the wrench brands now recorded can be traced to the one that was recorded to Miller and Lux, as the location of the brand is not given. Since we can record the same iron in eight different locations on cattle, the location of the design makes the recorded brand significant.

"The 'FG' connected brand, (once used by Pete French and Dr. Glenn) is now recorded to Allied Land and Livestock." 1972.

According to Idaho's Brand Recorder, "One of the oldest brands registered in good standing is the "J," left ribs on cattle and horses, left shoulder. It was first recorded on October 21, 1905 to a John L. Thomas of Malad, Idaho. Through the years it has been transferred to various stockmen but is presently recorded to an Edward Davis of Malad."

The "Ox Yoke T" (rib) and O (hip) brand was used on the first trail herd Nelson Story brought from Texas to Montana in 1866. Story recorded the brand in January of 1880.

Poindexter and Orr of Beaverhead County registered the Square and Compass, the first brand to be officially recorded in the Territory of Montana, probably in 1870 or 1871.

To handle the recording of brands, control theft and benefit animal health, the Montana Livestock Commission was created in 1885 under the name

Brands enough for a two year old. L.A. Huffman photo.
Courtesy Montana Historical Society, Helena, Montana

Board of Stock Commissioners. The first meeting was held at Miles City, Montana during which Granville Stewart was elected president.

The minutes of meeting called upon the Board to communicate with the U.S. Secretary of Interior and the Commissioner of Indian Affairs to authorize inspectors and detectives to enter Indian Reservations in the territory in search of livestock belonging to citizens and, if any were found, that the Indian Agent in Charge note the evidence and should it appear the stock was stolen "he shall turn over and give possession of said stolen stock to the Territorial Board of Stock Commissioners, inspectors, or detectives, or to the owners thereof, without necessary delay."

And that was not an unusual mandate to come from Montana's territorial cattlemen who were inclined toward prompt, decisive action.

Though "slick-earing" was viewed with some tolerance as a "necessary evil" in Montana and other territories, "rustling" was looked upon with "dead" seriousness.

"Milliron" roundup — Eastern Montana. Circa 1890. L.A. Huffman photo.
Courtesy Montana Historical Society, Helena, Montana

"Wrench" — CSC, Miller and Lux Brand.

Rustlers lost more than reputation when caught. Vigilante action left quite a few rustlers dangling at the end of a rope.

The effectiveness of Vigilantes, no doubt, was an inspiration to cattlemen to join together in solving theft and other problems common to the raising of cattle.

During those early years of violence, British Columbians felt the "Mounties" kept lawlessness under better control than the deputies and marshals to the south.

Cattle rustling has continued sporadically throughout the years in spite of more rigid inspection of both live and slaughtered beef during transportation.

An article in the Walla Walla Union Bulletin dated April 10, 1951, expressed the concern of Umatilla County Cattlemen who "put their heads together one day last week and discovered at least 50 good steers had been snitched by rustlers in that county since January. President Roy Duff and his colleagues launched a resolute campaign to stop this costly nuisance. The men offer $500 for evidence leading to conviction of an act which not too many years ago got summary punishment in the form of hanging under a fence-rail tripod, if no tree was convenient."

From time to time, changes have been made in livestock regulatory measures to keep pace with a growing industry in an increasingly more complex environment.

Methods of branding have changed little, although alternative ways have become available for cattle-

Branding irons have been electrified, eliminating a need for a fire where branding can be done near a source of electricity.

A method of chemical (freeze) branding was devised in the 1960's but has not been widely accepted and is not considered legal in some areas except for on-the-ranch identification purposes.

Different systems have come under consideration. One system proposes a symbol of lines that can be arranged in an almost infinite number of brands and would have application for cattlemen on an international basis.

Another utilizes a coded capsule which can be implanted in an animal and 'read' with a scanning device.

As good as these and other ideas may be, replacing the old "hot iron" method will be difficult.

Present-day ranchers, many of them descendents of pioneer cattle families who registered the first brands, prefer to use the brand of their fathers.

The feeling he has as he holds his branding iron in his hand and puts it on a quality beef animal will be hard to forget.

"Rounding up 1200 head of cattle into the corral at the dipping vat. The next day the boys were branding calves and, while at dinner, a prairie fire started from the branding irons. The cook went ahead and backfired and the men dragged a steer along the edge of the fire and put it out. Saved thousands of acres of fine food. The cook had never cooked on the range before but had read in papers how to backfire." G.V. Barker, photographer.
Courtesy Montana Historical Society, Helena, Montana

146

An old time tailhold — L.A. Huffman photograph.
Courtesy Montana Historical Society, Helena, Montana

A branding scene. Greenleaf cattle in the Miller Coulee, Rosebud County, Montana in 1920.
Courtesy Evan McRae

Branding an "XIT" whiteface near Cedar Creek in 1910,
Cowboys are: left to right: Warren Johnson, Pete Sherry,
J.K. Marsh, Bill Foaght.
Courtesy Historical Society, Helena, Montana

Lee Fleegle, boss of the Prouty
Commission Cattle Co. at the
"Pee Muleshoe" Ranch, Valley
County, Montana in 1910.
*Courtesy Montana Historical
Society, Helena, Montana*

Branding scene.
*Courtesy Utah State
Historical Society*

Branding in a chute. L.A. Huffman photo.
Courtesy Montana Historical Society, Helena, Montana

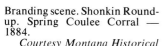

Branding scene. Shonkin Round-up. Spring Coulee Corral — 1884.
Courtesy Montana Historical Society, Helena, Montana

Woodard Brothers branding calves at Nicola Valley, British Columbia. Circa 1895.
Courtesy Provincial Archives, Victoria, B.C.

Branding on the Lower Judith
River — Circa 1910.
*Courtesy Montana Historical
Society, Helena, Montana*

The Oblate Fathers who founded and maintained the
Cariboo Indian Mission have assumed many of the crafts
of the cowboy down through the years and now are as
professional in many ways as the full time Cariboo cowboy.
One such man is Father Larkin seen here applying a brand.
Courtesy Williams Lake Tribune

Crow Indian cowboys branding for the I.D. Company herd about 1890.
Courtesy Montana Historical Society, Helena, Montana

Branding cattle at Kershaw.
Courtesy Montana Historical Society, Helena, Montana

Early squeeze chute on Gordon Ranch on Asotin Creek, 1918. Dave Martin, Floyd Gordon and Clyde Gordon.
Courtesy R.P. Weatherly

Lewis Campbell ranch scene. 15 miles east of Kamloops, B.C. Campbell was one of the enterprising individuals reknowned for making rigorous, long trail drives to get beef north to the miners. He settled down to develop this ranch. Here Campbell cowboys are at work. The rider on the left is Alvie Shafer; Lew Campbell in angora chaps; rider on right is Walter Campbell. Photo taken around 1900.

Courtesy Provincial Archives, Victoria, B.C.

Cattle being branded in Idaho.
*Courtesy Idaho State
Historical Society, Boise, Idaho*

Branding on the "Figure 3" Ranch. A scene that has occurred on this site for more than 100 years. Sprague, Washington.
Courtesy Clarence Dooly

Ranch Branding Scene: Joe Kilsey, Valley Rinker, Ross Woodard. Tending fire: Paul Lowden, Arney Will mounted.
Courtesy Ross Woodard

Putting on the "Figure 2."

Courtesy Ross Woodard

Cows waiting for calves to be branded. Ranch corral scene.
Courtesy Lawrence F. Bradbury

Round up and branding at Dick Ranch, 1940.
Courtesy R.P. Weatherly

Branding a "critter" on the Bones Brothers Ranch near Birney, Montana.

Courtesy Burlington Northern

There are few innovations to making roping a calf any easier than it was in the 1950's or 1850's. Man and horse work as a unit, with the man tossing the rope and after the calf is lassoed, tieing the rope to the saddle. The trained, cowboy mount will hold the rope taut. Don Forest, Asotin County, Washington and Dr. Jack Raaf, Portland, Oregon, did the cowboying while Byden Tippett and Jack Raaf, Jr. worked on the calf.

Branding at Cactus Flat Corral, Tippet Ranch in 1952. (Jack Tippet, center) Southeastern Washington.

Close up of branding operation which only takes a few minutes. Calves have been separated from the mother cow and wait in the background for their turn under the hands of men skilled in application of the hot iron.

"Cactus Flat" corral branding scene along Grand Ronde River.

Courtesy R.P. Weatherly

Two ropers at work heeling 'em during branding at Tuff Webster's ranch.

Courtesy Northwest Unit Farm Magazines

A roper in the process of taking a calf by the hind legs, a skill not learned overnight.

Courtesy Northwest Farm Magazines

"Hind Legging" a doggie near Glasgow, Montana. Circa 1950. Bill Browning photo.
Courtesy Montana Historical Society, Helena, Montana

Branding on the 'Top Hat' Ranch near Twodot, Montana.
Courtesy Montana Historical Society

Close enough to smell the smoke.
Courtesy R. P. Weatherly

Picking out the calves. A Montana cowhand drags a roped 'doggie' to the branding iron on a Montana ranch near Hinsdale in far eastern Montana. "Hind Footin'." Bill Browning photograph.

Courtesy Montana Historical Society, Helena, Montana

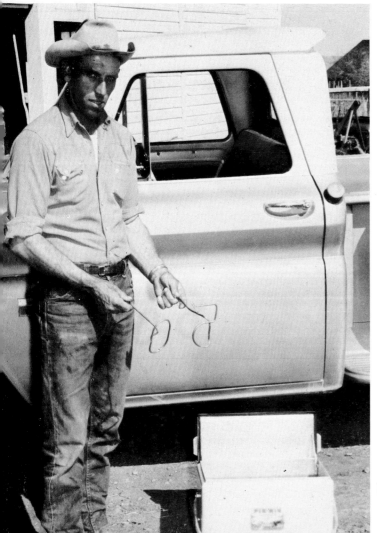

Jim McDonald, Idaho County, Idaho displays the copper irons and equipment he used to apply freeze brand.
Courtesy Northwest Unit Farm Magazines

Placing the heated iron on an animal securely held by a calf table. The table tilts making the task easier.

156

Dr. Landers places a copper freeze branding iron on a Hereford cow at the Beale Stock Farm, Washington County, Oregon. Long hair was clipped as short as possible and the skin was wet thoroughly with alcohol before application of the iron. Oregon regulations require that within herd identification be placed on either shoulder and consist of at least three numbers or letters or a combination of each. It is recommended that the first number represent the year of birth; i.e. 701 for the first born in 1967.

Photo by Vern Lewis.

Courtesy Northwest Unit Farm Magazines

A good example of freeze branding. The brand is a Connected HB Quarter Circle.

Courtesy Northwest Unit Farm Magazines

Six hundred cattle being moved from Asotin County, Washington to summer ranges in Oregon. The "Saddle Knife" Brand shows on the right shoulder of the animals for positive ownership identification.

USDA Soil Conservation photograph by C. Dillon.

CHAPTER VIII
Builders of Beef

The foundation of the Northwest cattle business is unique. The industry began with a nucleus of longhorned Spanish cattle, descendants of Old World Bovines, long associated with man. In the 1840's, Dr. John McLoughlin, Hudson Bay Fur Company's Columbia District Factor, added strains of "Durhams" for better milk and meat production. Descendants of European cattle brought to the New World in the 17th Century and Durhams, imported to the Colonies from England in 1783, were brought West by settlers. Cattle commerce in the new land, for reproduction and expansion purposes, intermingled these cattle of different breeds, creating a variety of color, size, conformation and quality.

In the colorful, turbulent days of Western Frontier expansion, more thought was given to taking advantage of natural forage on open range, than to herd improvement. Nor was there a particular need. Cattle population became so vast that few ranchers knew the exact number of head they owned. Cattle grazed at will in great numbers on an empire of grass, an empire that for centuries had provided food for native animals including the buffalo.

Thousands of acres, available by occupation, offered opportunities, seemingly unlimited, and invited accompanying abuse. Investors were attracted from all over the continent and Europe, as well. Some were successful and others failed.

Another aspect of the beef industry in the early days was the entry from the Southwest of large droves of Texas Longhorns to fatten on free grass mainly for shipment by rail to the East.

Overcrowding of ranges, death loss caused by weather, predators and disease, loss of land to crop cultivation and growing population encouraged herd improvement as a means of raising more beef.

Although some effort had been made by more visionary individuals earlier to upgrade cattle in the Northwest, it was not until the late 1800's that a concerted effort was made toward herd improvement.

A Klickitat County, Washington cattleman describes his cattle operation in 1880 which is typical of ranching in the vast territory and of the times.

". . . number of cattle owned, January 1879, 5000; number of calves branded during 1879, 1,725; number of cattle sold during 1879, 500. Twelve thousand five hundred square miles of free range government land; 160 acres of land owned.

The average composition of a herd of 5,000 just before round-up was: bulls, 100, average value of each, $15; cows, 2,200 average value each, $12; steers, three years and upward, 500, average each, $20; two-year olds, both sexes, 900, average value each, $12; yearlings, 1,300, average value each, $8. Steers in great demand, and hence a small proportion are held by cattle-raisers. One-half of the yearlings and two-year-olds are males. They run so near even that many stockmen keep the talley together and divide by two; one year with another it is an even thing. If properly cared for, a herd of cows will bring 75 per cent increase.

". . . One thousand head is the smallest number that can be run here. About 40 bulls to 1,000 cows are allowed. (Cattle belonging to a number of owners occupied the same range and guide rules were set up. Author) The high-bred bull should be brought in not older than a yearling, but does not get fully acclimated and useful until the second generation.

"The weight and quality of cattle have improved by the importation of Shorthorn bulls in about the same proportion as the ranges have decreased; if the stock had not been improved they would have decreased in weight and quality in the same proportion as the ranges filled and began to fail. Shorthorn bulls have been brought into the country and renewed often. Their progeny have helped improve our common stock. Shorthorn cattle will be run in small bands, and taken better care of. Running cattle in large bands in this country is fast getting to be a thing of the past."

The Texas Longhorn is regarded as a product of the North American Continent, a result of natural selection and adaptation to environment. Its ancestry can be traced to the black Spanish fighting bulls, descendants of cattle driven into Spain by the Moors, to the extinct aurochs, antecedants of all European cattle. When mixed with straying, domestic cattle from wagon trains and settlements, the hardy, active animal evolved.

Courtesy Texas Longhorn Association.

Red Poll cattle in left foreground. Picture taken in 1870. Red Poll cattle originated in the East of England during the first half of the 19th Century, with the merging of two native stocks, the Norfolk and Suffolk, descendants of red and polled cattle that existed on the European continent since the early times of recorded history. The breed has been in America as a registered breed since 1874.

Courtesy Porter Griswold

159

A Yakima County cattleman submitted the following notations for publication in the Tenth Agricultural Census, June 1, 1880, Washington, D.C.:

"The way cattle are handled on our range there is no use in a man having less than 1,000 head, unless he runs a dairy in connection with the other parts of the business. About 30 bulls are allowed to 1,000 cows. The high-bred bull should not be older than two years when brought here. Their half-breed descendants are as hardy as any of our cattle."

This report and the preceding one, indicate the efforts of ranchers to increase productivity by improving the quality of their cattle.

Most "high-bred" bulls were secured from growers in the Eastern part of the United States or imported directly from England. The business of raising pure-breds was being slowly implemented in the Northwest. Within 20 years great strides had been made in taking advantage of the opportunity offered by commercial herd operators for a local supply of higher grade breeding stock.

Experts from the Red Poll American Herd Book, Volume II published in 1890 (a joint enterprise with the British Society) reveals the growth of purebred enterprises:

"John Blurock, Vancouver, Washington is credited with the registration of one bull, 'Washington's Duke 2130.' " Other entries from the Northwest were L.K. Cogswell of Chehalis, Washington, 3 bulls and two females; John H. Moore, East Portland, Oregon, one bull and D.M. Ross, Puyallup, Washington, one female.

In volume 22, 1901, registered in the first independent American Red Poll Herd Book (previously American Red Polls were registered in the British Herd Book), the entries substantiate significant expansion:

Pheonicians may have brought ancestors of Devon cattle to Southwest England and these red cattle were noted by the Normans when they invaded the area in 55 B.C. The breed is named for the County Devon where it has flourished for centuries. America's first meat packer, William Pynchon, founder of Springfield, Massachusetts made corned beef from Devon cattle in the 1640's.
Courtesy American Devon Cattle, Inc.

General Members: Robert G. Barkley, West Home, B.C., Canada; Wm. Bryan, Canyon City, Oregon; A.D. Clark, Boise, Idaho; C.T. Gibbons, Cowichan Station, Vancouver, B.C.; E.A. Hinkle, Roseburg, Oregon; Geo. Lazelle, Oregon City, Oregon and F.H. Porter, Halsey, Oregon.

Herds and Owners: T.D. Allen, Silverton, Oregon; T.W. Baker, Elgin, Washington; Wm. Bryan, Canyon City, Oregon; Mr. Cogswell, Chehalis, Washington; J.H. Grace, Central Point, Oregon; E.A. Hinkle, Roseburg, Oregon; J.T. Hinkle, Roseburg, Oregon; C.L. Holmes, Albany, Oregon; E.W. Mahoney, Tekoa, Washington; John McDonald, Mt. Vernon, Oregon; Alexander McPherson, Twin Falls, Idaho; F.H. Porter, Halsey, Oregon; Porter & Curtis, Halsey, Oregon; and W.L. Styers, Miner, Montana.

Red Polls were not the only breed being developed in the Northwest. The foregoing record was selected as representative of breed growth.

Though commercial herds provided beef for consumption, production has been bolstered by the devotion and loyalty of individual cattle families to a particular breed.

Significant of this deep sentiment, as of June 1966, sixty-two lifetime and annual memberships in the Scotch Highland Breeder's Association were listed from the states of Oregon, Idaho, Washington and Montana.

In the late 40's Shoop Brothers of Browning, Montana had one of the largest herds in the country. Many present-day herds were built on foundation stock from the herd of Cyrus and Jess Shoop who are credited with preserving Scotch Highland stock in the North American continent.

They built up a herd of several hundred cows from purchases of cattle brought to Livingston, Montana from Scotland by Walter Hill who at one time, owned more than 300 head of Scotland's finest Highland cattle.

An American Milking Shorthorn Society was formed in 1920, the chief objectives of which were to bring about closer cooperation in breeding true dual-purpose cattle and to promote the breed, both horned and polled.

American Colonists relied on the old, red cow descending from England's Devon bloodlines and these cattle were brought along with wagon trains of immigrants to the Northwest becoming an integral part of the cattle industry.

In later direct importations, James J. Hill, Great Northern Railroad magnate, is credited with bringing a load of pure-bred South Devon cattle to America in 1915. The cattle were distributed in the states of Washington, Idaho, Montana, North Dakota and Minnesota.

Devonacres in Eagle Point, Oregon is the site of one of America's finest herds of Devon cattle. Started as a hobby project, the herd has increased to a working operation of more than 200 head.

Though the disposition of Brahman cattle is often questioned because of a wild, rodeo image,

160

Polled Shorthorns
Naturally polled Shorthorns were the foundation of the Polled Shorthorn Breed, including the famous twins "Mollie and Nellie Gwynne" part of a small herd belonging to the McNair Estate near Minneapolis, Minnesota. The herd was first reported in 1888.

Shorthorns
Shorthorn cattle originated in Northeastern England in the Valley of the Tees River. During the early days of importation to North America, these cattle were referred to as "Durham Cattle."

A Rabinsky photo

the Brahman Breeders maintain the breed is intelligent, inquisitive and can become very docile. Mrs. Tom Colvin of Ritter, Oregon is among breeders and advocates in the Western United States. Though the breed's greatest popularity lies in the Southwest, where an organization was formed in 1924, western cattle feeders have been impressed with the gainability of Brahman hybrids particularly during hot, summer months.

Another hybrid is the Brangus, a combination of European Angus and the Asian Brahman. The first Brangus breeder in Washington was Lloyd Daily of Kennewick although the honor of being first in the West belongs to Ed Shafer of Walla Walla, Washington, formerly of California. Oregon's first can be claimed by Lacomb Brangus Farm at La-Comb. An active Idaho breeder is Howard Willson of Emmett. A big Timber, Montana man, Lyle K. Jones and a Farmington, B.C. grower, Allen J. Low are among active "Brangusmen."

Charolais is an important French breed and has gained recognition in the area as a beef improver. A recent report of a Montana crossbreeding program points to advantages of heavier weaning weights and feedlot gains for Charolais crossbreeds.

Early days Asotin County cattleman, Clyde Gordon, with a Durham bull. Photo taken at Cloverland, Washington in 1910.

Courtesy R.P. Weatherly

Bright Hope, high-quality Hereford cow shown by A. J. "Jack" Splawn, owner, and his herdsman, Norman Hale (holding whip). Cattle occupied much of Splawn's life from slow moving oxen on the immigrant trails, early droves of beef to distant markets to a foundation herd of English-bred Herefords. With his wife, Margaret, the daughter of a pioneer family, he built a successful purebred enterprise that influenced the entire Pacific Coast cattle industry. The names of Splawn Hereford lines lie buried in hundreds of western pedigrees.

Courtesy of Homer Splawn

161

These cattle are of the oldest Hereford herd in continuous U.S. registry. Part of the Beau and Belle Donald Herefords are shown on the 1,000 acre Curtice Martin Ranch. The ranch is on the east slope of the Bitter Root range of the Montana Rockies.

Courtesy of Burlington Northern

Following the early introduction of Shorthorns into Northwest cattle operations, a most significant impact was made and has been maintained by the Hereford Breed. The breed's acceptance and performance was quite dramatic and it is still a highly popular breed.

Jack Splawn, pioneer cattleman and drover, bought a Hereford bull calf at Indianapolis, the first sale of English Imports held that far West, around 1887. The bull, purchased for $1500, became a part of his foundation herd from which offspring were exported to the Sandwich Islands, California, Alaska and to the Manchu Empress of China. Splawn's herd was the first purebred Hereford operation to be established in Washington state.

The Aberdeen Angus have also retained a predominate role among herds in British Columbia and the Northwestern states. Among firsts in the beef business, the American Angus Association offered a production records program, providing an invaluable tool to increase production efficiency.

A kin to Aberdeen Angus is the Red Angus, a pure breed and one of the latest to be formally recognized in the United States.

Today's program of beef breeding is largely a matter of individual choice and experimentation under controlled circumstances.

Whatever the breed and importance attached to it, producers are interested in high production efficiency as population in the area becomes more

The *Polled Hereford* differs in only one way from the Hereford, it is hornless. The breed sprang from mutations occurring occasionally among purebred herds. The breed was enthusiastically promoted after discovering that polled bulls would sire polled calves when mated to horned cows. The breed rapidly increased in popularity and spread to western ranches in the 1930's.

dense, land values and taxes increase and good land diminishes. Modern producers are looking for answers in confinement systems, artificial breeding, multiple births and different breeds with a focus on "exotics."

There is interest in the Blonde d'Aquitine, Limousin, and Main Anjou from France; Fleckvie and Gelbvieh from Germany; Chianina from Italy; Simmental from Switzerland; Murray Gray from Australia; Galloway, Lincoln Red, Welsh Black from Great Britain and Hays Converter from Canada.

Such names as Barzona, Beefmaster and the Pan American Zebu are contained in experimental reports along with dairy breeds Brown Swiss, Jersey and Holsteins.

And the Texas Longhorn — will this extraordinary animal again be a part of the Northwestern range scene? Texas Longhorn Association proponent Walter B. Scott believes so:

"Our main purpose is to maintain and preserve this magnificent breed of cattle . . . European breeds need careful watchfulness on a range when temperatures are high . . . such is not the case with the "original Texas Ranger" . . . Texas Longhorn Cattle are not God's gift to the cattle industry but damn sure were a blessing to Texas and the Western cattle industry in the beginning. We sincerely believe they still have a lot to offer in restoring new vigor, hardiness and "help-make-it-on-their-own" ability."

Pictures show conformation (shape) changes within the last twenty years.

The ideal beef animal of today is not the lean, sinewy animal of the free and open range nor is it the boxy, plump pure-bred exhibited in mid-20th Century shows.

It is a highly improved version that now falls in between the two extremes.

Consumer demands for a more lean tender meat, increasing population, growing per capita consumption and keen competition with other foods at the market place, present an intriguing challenge to Northwest builders of beef.

The Hereford trade-mark is a white face accompanied by a red body and white markings on legs and crest or lower line. It is the product of excellent breeding practices in England's Valleys of the Severn and the Wye. Survival of hard Northwest winters established the reputation of Herefords and vast numbers were imported from England. Many fine herds of "seed stock," antecedants of these cattle, exist throughout the area.

Courtesy of American Hereford Association

The *Galloway* is closely related to the Aberdeen-Angus and is native to a district, Galloway in extreme Southwestern Scotland. The Galloway is not as wide of back nor as deep of body as an Angus but it produces a high quality carcass. The Galloway has a wonderful coat of long, black or brown-black hair consisting of two parts, a silky undercovering and a long, soft, curly outer coat, especially suiting the breed to cold, winter exposure.

Aberdeen-Angus cattle are native to the highlands of Aberdeenshire and Angusshire, Scotland. Reference was made to black polled or hummel cattle in 1523. There is substantial evidence they grazed the hills of Northern Scotland centuries before. Red Angus have the same background, genetic makeup and bloodlines of the Black Aberdeen-Angus. A significant exception is in color. The Red Angus Association was organized in 1954.

Courtesy of American-Angus Association

Scotch Highland
In the first Herd Book published in Scotland in 1885, the question of whether or not Highland cattle were an aboriginal breed was posed. It was determined that if the Wild Cattle of Chillingham Park (Scotland) were representative of aboriginals, the Highlanders were also. This breed originated on the Mainland and Western Isles of Scotland. There were two distinct classes, the West Highlander, Kyloes, and Highlander or Main Highlander. Kyloes were smaller, shaggier and mostly blacks.
Courtesy of the Scotch Highland Breeders Association

The American Brahman breed originated from bulls and females of several strains of Indian humped cattle brought to the United States between 1854 and 1926. Because the Brahman has the ability to withstand extreme heat and is resistant to insects, it has made a valuable contribution to the beef industry. The size of the animal plus these qualities has encouraged crossbreeding programs and led to the development of the Santa Gertrudis, fusing the bloodlines of Brahman and Shorthorn. Other crosses include the Brangus (Brahman with Angus), Beefmaster (Brahman with Hereford and Shorthorn) and Braford (Brahman with Hereford).
Courtesy of the American Brahman Breeders Association

One of the oldest and highly popular breeds of French cattle, the *Charolais* is considered to be of Jurassic origin and was developed in the district around Charolles in Central France. There is historical evidence that the white cattle were being noticed as early as 878 A.D.
Courtesy American-International Charolais Association

The Canadian Cow
In 1610, Jacques Cartier brought these hardy, brown-black cattle to the North American Continent from Brittany and Normandy, Northern France. By the 1700's, there were French cattle at the fur-trading posts and missions along the St. Lawrence River in Eastern Canada. By the 1860's, the Canadian Cow was well-recognized for hardiness and ability to produce milk. A French-Canadian cattle breeders association was formed in 1895.
Courtesy Williams Lake Tribune

Dexter
Lee's Hill Liberty Belle, Number 1007. Owned by Marvin A. Mackey, Vancouver, Washington

Dexter cattle are a variety of the Kerry breed originating long ago in Ireland. It is believed the North Devon cattle were taken to Southern Ireland and bred to Kerry stock. Mr. Mackey, a Dexter fancier for fifteen years, finds the breed "easy to keep and good for both milk and beef."
Courtesy Marvin A. Mackey

Yearling bulls from cooperating breeders' production-
testing program are inspected at the end of the 1954 test
year at the Caldwell Branch Station, Idaho. In the background can be seen the individual feeding facilities.

Courtesy Northwest Unit Farm Magazines

From the days of the Vigilantes to the present, cattlemen have united to solve problems of
cattle raising — "to do for a cattleman what he cannot do alone." Associations are active
in British Columbia, Idaho, Oregon, Montana, and Washington. 1936 cattlemen's meeting
at the Beacher House on Whiskey Creek near Chilcotin Mountains in British Columbia.
Seated left to right: Barney Baron, Dave Melville, Dr. Dunn, Department of Agriculture,
Wallace MacMoran, manager of the Gang Ranch, Professor Bucklee, etomologist, not identified,
"Pudge" Moon. Standing left to right: Forestry official, Corporal Gallager, George Renner,
Roderick McKenzie, Mrs. F.M. Beacher, Premier Talmi, Mr. Beacher, Melville Moon, Minster
of Agriculture Munro, Bill Muir, Talmi's Chauffeur, R.C. Cotton, Department of Agriculture
representative seated on the fence.

Courtesy R.A. Moon

CHAPTER IX
From Feed to Finish

Back in the old days, few cattlemen harvested forage for their livestock.

In the vernacular of an early rangeman "Some parties prepare hay for feeding in winter, many do not. Those who do not have hitherto lost as few cattle as those who do."

As long as grass grew under Old Bos' hooves, it was converted into meat and milk.

But grass became less plentiful as time passed. Free range grass became "too short to make it profitable to own cattle without a ton of hay to every cow."

More changes took place in "putting up" hay and feed in the first half of the Twentieth Century than in all the centuries gone past.

Hay cut for feed in the late 1800's was wild-grass hay. It was cut with a mowing machine, an implement pulled through the meadows by a team of horses or mules. As the wheels turned, gears operated a bar equipped with saw-edged teeth which clipped the grass.

The grass was gathered into piles by hand. Later, a horse-drawn hay rake replaced hand labor. Half-circle tines raked the grass as the machine was pulled forward. The operator released the load at regular intervals along the way.

The hay was allowed to cure for a day or two, depending on the amount of moisture in the air. If the weather was extremely hot and dry, the process required less time. Grass was harvested in the summer at the right peak of maturity.

After the grass had been mowed and raked or gathered into small piles (shocks), it was pushed by a horse-drawn 'buck' to a hay stacker which lifted the loads of loose grass and dumped the bunches into a larger pile.

Using pitchforks, men shaped the forage, layer upon layer, into stacks. It was essential the hay be "stacked" properly to avoid spoilage caused by weather and to keep the stack intact.

There were risks involved. In rattlesnake country, it was not uncommon for a snake to be gathered up with a load. Or, a cable on the stacker might break, whipping through the air or allowing the fork to fall on an unwary worker below. Putting up hay was hot, dusty work.

When clouds appeared on the horizon, long hours were spent in the hay fields to finish the task before the downpour. Rain ruined many crops, causing spoilage and originating the saying "Make hay while the sun shines."

Yields were one-and-one-half tons to the acre. When winters were mild, the stores of hay accumulated, enabling more cattle to be held during long, cold seasons.

If barns were available for storage, some hay was brought in from the fields by team and wagon to be lifted into a haymow, a favorite place of ranch youngsters to lay and dream of days when they would have a ranch of their own.

In the early 1900's, a stationary hay baler was used to facilitate the handling of hay. By modern standards, it was a monster.

The machine was pulled to a stack of hay where it compressed loose hay into rectangular blocks or bales.

Bales were wired by hand. Two men stood by the press to wire each bale as it came through the machine.

Stationary baling on ranches was custom work. The owner of the baler hired his own crew of from twelve to fifteen men, moving from ranch to ranch throughout the area.

Nearly fifty years elapsed before a smaller, more compact machine began to appear on the market. Early models were pulled behind a tractor and tied the bales automatically.

Concurrent with improved baling machines, new mowing and raking concepts appeared on the ranch

A Bulger grass-fed range steer. The steer weighed 1840 pounds at two years of age; 3230 pounds at eight years.

Courtesy Idaho State Historical Society, Boise, Idaho

scene, making further reductions in manual labor and conditioning the hay to retain utmost nutritive value.

Within a few years, improvements were made incorporating many haying operations into one. Equipment became available to mow, condition, chop and transport forage to on-the-ranch storage facilities. Hay could be cubed, pelleted or baled in a broad variety of shapes and sizes for easier handling.

The day may come when manual labor will be removed from the haying process.

As changes were made in preparing feed, cattlemen became more efficient in its utilization, producing more pounds of meat for pounds of feed.

Cattle consume about fifty percent roughage, material that is not edible by humans. The other half consists of grains which have been a part of the human diet for centuries.

From the earliest times of recorded history to the early 1900's, cattle were fattened by letting them graze on natural grass and meadow land.

Steers were four and five years old before their weight and size was sufficient to produce a marketable carcass. By supplementing the animal's diet with grain and nutrients, the time required to produce market-ready cattle was reduced.

From the 1920's to the 1940's, the number of grass-fed steers gradually diminished. At present, few cattle go to market without being fed a nutritionally balanced ration.

During this time, few cattlemen would admit to feeding dairy-beef — beef derived from cattle or dairy breeding or from dairy and beef crossbreds.

With the development of confined cattle feeding and weight-gain record keeping, came the realization that dual-purpose cattle did not belong to the ages.

There was a time when Holsteins were known as "magpies" and Jerseys as "yellow hammers." The meat packer was hesitant to show visitors the carcasses of these animals because of the yellow fat. When fat was yellow in color, it indicated the meat had come from a dairy breed or that the animal had been fattened on grass.

Today's cattlemen are more concerned with rate of growth, feed efficiency and net returns.

Formerly, dairy cattle were sold for veal. Now more and more dairy steers are being routed to the feedlot. Almost one-third of the beef consumed is dairy-beef.

Changes in feeding methods brought changes to the marketing route from ranch to retail counter.

170

Instead of direct-to-packer from the range, a new marketing pattern emerged.

From the cow-calf operator on the ranch, cattle were transported to marketing centers, where the cattle were purchased by the feedlot operator.

One of the most intriguing aspects of cattle trade occur at the sales yard where cattle are auctioned.

When the auctioneer steps onto the auction block, the audience becomes silent.

The auctioneer is a showman, taking advantage of the buyer's eagerness to obtain an animal at bargain price while the seller is hoping to "top the market," — receive the highest price per pound at the sale.

After describing the animal, or animals, in the ring, the auctioneer slides into his chant, opening the bidding near the current market price.

Bidding clues come in many forms. A bidder may tip his hat, wink, scratch his nose or nod his head. Unless an individual is interested in purchasing cattle, he is wise to refrain from waving at a friend, brushing at a fly or fanning himself with a paper on a hot afternoon.

When the gavel raps "going, going, gone" a beef animal is another step on the way to the meat packer and wholesaler. More than a century has passed since beef was butchered "on the spot" and sold at early mining camps. Meat packing has become a sophisticated, sanitary, highly competitive business involving around thirty processes from live animal to dressed carcass.

Beef carcasses are boned, trimmed and portioned at the meat packing plant. The trend is to increase the services at the processing level and reduce those at the retail meat store.

After the primal cuts are delivered to the retail meat market, the meat is made into specific cuts of beef, trimmed, packaged, weighed and labeled for the convenience of the consumer.

Consumers of today are seldom acquainted with the beef producer — breeder through feeder.

Distance makes personal contact impossible in the Northwest, and elsewhere, as cities grow larger and agricultural communities shrink.

The future of human survival depends upon producing more food for more people in a shorter period of time on less land.

Those who defend converting grass to beef have a strong case. Milk and meat are the best sources of high-quality protein, the basis of all life.

A conservative estimate of lands suited only to grazing purposes would be twenty acres of land per animal . . . or nearly half of the total land in the states of Washington, Montana, Idaho, Oregon and the Province of British Columbia. There are more than 8 million "four-legged" protein-producing factories converting this grassland to food and fibre for public consumption.

Old Bos and the cattleman have contributed significantly to the thriving, complex world of America's Northwest.

Early cattle feeding scene taken in the Boise Valley, Idaho.
Courtesy Idaho State Historical Society, Boise, Idaho

Putting up hay without horses was a dream for the future. This picture, taken in the early 1900's at the E. & I. Roper Farm, Burnside District, near Victoria, B.C., shows draft horses hauling hay to the barn for storage.
Courtesy Provincial Archives, Victoria, B.C.

Six-horse teams were used in 1900 to haul hay and grain from Bingen to White Salmon, Washington. Teunis Wyers was the owner of the horses.
Courtesy Mr. and Mrs. Russell Kreps

Granny Tillman and Cecil Hutchings haying in the 1920's Okanogan County, Washington.
Courtesy Ross Woodard

Before the railroad reached the Klickitat County area, feed grain, freight, and supplies were shipped by river boat. One of the Columbia River Transportation boats on its regular run from Portland to White Salmon at the turn of the Nineteenth Century.
Courtesy Mr. and Mrs. Russell Kreps

Johnny Lauderbach's crew baling hay in the early 1900's near Bingen, Washington, his father came to the area from Texas in 1892.
Courtesy Mr. and Mrs. Russell R. Kreps

Stacking hay on a farm at Pemberton, British Columbia.
Courtesy Provincial Archives, Victoria, B.C.

Stacking hay on the G-D Ranch in Montana's Big Hole Basin near Dillon.
Courtesy Montana Historical Society, Helena, Montana

Archdale Ranch, Knowlton, Montana, winter feeding scene
Courtesy Montana Historical Society, Helena, Montana

Stacks of hay for winter feed. "BX" Ranch (F. S. Barnard) at Vernon, B.C.
Courtesy Provincial Archives, Victoria, B.C.

Hauling hay at the Dick Ranch, Asotin County, Washington in 1935.
Courtesy R.P. Weatherly

Feeding during a snow storm. Alma Z. Clausen at Roundup, Montana in 1949.
Courtesy Montana Historical Society, Helena, Montana

A demonstration of hay baling equipment.
Courtesy Northwest Unit Farm Magazines.

From hand-stacking to team and wagon hauling, to mechanized baling and auto freight, forage preservation and storage have forged ahead just as other phases related to beef production keep pace with scientific and engineering advancement.

Charles Gibson cattle in range corrals on the Cowiche. Gibson owned two Townships in the area.
Courtesy Charles Gibson.

Greenleaf Cattle Company stock ready for shipment to St. Paul, Minnesota. The cattle were loaded at Colstrip, Montana which is twelve miles from the ranch. 1935.
Courtesy Evan McRae

Cattle corrals at Williams Lake, B.C. in 1947.
Courtesy Provincial Archives, Victoria, B.C.

Part of the Pacific, Great Eastern Stockyards at Williams Lake, B.C. Cattle being held for shipment.
Courtesy Provincial Archives, Victoria, B.C.

Shipping beef from the ranges in Montana to the Chicago, Illinois market.
Courtesy Montana Historical Society, Helena, Montana

Following ranch sale, the cattle are penned for shipment by rail. The animals in the corral are polled and uniform. Circa 1950.

Midwest marketing center operating at fairly high level of capacity. Changes in transportation and relocation of packing plants closer to the source of cattle have changed the marketing pattern. Some terminals have closed such as the reknowned Chicago Livestock Market. Circa 1935.

By 1947, grass fat steers were almost a thing of the past. More marbling and higher quality beef could be produced by confinement feeding of a balanced ration of Northwest grains. Research has been conducted to achieve the best ratio of feed to pound of meat produced. The age of computer processed data has taken the guess work out of feeding for efficiency.

176

Mr. Charles H. Frye — Early Meat Packer Charles Frye came West from Iowa in 1884. He established a small meat shop in Montana, expanding into the range and cattle business in that and other areas. In 1888, Frye moved to Seattle, Washington where he opened a number of meat markets, one financed with borrowed capital. The note was secured by a chattel mortgage on "one large platform spring meat wagon, one bay or brown saddle horse six-years-old, one saddle and bridle for $345.00." He and his associates erected a meat packing plant in 1891 with a capacity of one carload of cattle per week. The plant grew to the largest single meat packing industry west of the Mississippi. He operated the plant until his death in 1940. Charles Frye operated company feedlots at Toppenish, Washington. The land was leased from Indian woman named Susie Spencer. Manager for the feedlot was Fred Fear, an early Yakima Valley stockman. Cattle pens were built to accommodate six or seven thousand head of feeders. Facilities included a cook house and bunkhouse for a dozen or more cowboys. The cattle were fed chopped alfalfa which was grown in the area. It was the only feed provided for the cattle in the early days of livestock feeding. Meat shipments were made to Hawaii and the Philippines via the Seattle plant which was served by train on a daily basis.

Courtesy Frye and Company.

FRYE & COMPANY PACKING PLANT

UNDER CONSTRUCTION . . . at the same location where Frye meats have produced for 55 years. The new plant will be of concrete construction interior and will be one of the finest in the nation.

A B-29 bomber crashed on the original plant in February of 1943 destroying the main building. A new plant was completed in 1944.

Courtesy Frye and Company

The Frye Packing Plant in 1890. At the time the area was all under tide water. An elevated streetcar track and interurban track passed in front of the plant. The land was hydraulically filled and is now the Seattle Tidelands.
Courtesy Frye and Company

Charles Gibson, an early meat packer, said "Cattle came into the plant from all directions." Gibson sent his cowboys to many points in Oregon and Washington. In reminiscing about the delivery of cattle, Gibson recalled, "Out in the hills between Walla Walla and Oregon, the cattle would graze all night. You could lay down and sleep — just dig a place out for your shoulders." "Cattle, even in the 1920's, were four-year-old grass cattle, gradually coming down to three-year-olds. They were mostly big, rough Shorthorns. Then settlers found they could raise corn in this part of the country and started feeding it in the 1940's. Gibson slaughtered "about thirty-five hundred animals per year, eighteen to twenty per day on the days when killing." He ran a direct-to-retail business, delivering meat to stores in horse-drawn meat wagons.

Photo by Author

The Yakima Meat Company in 1920. Center right is the old meat packing house.

Courtesy Charles Gibson

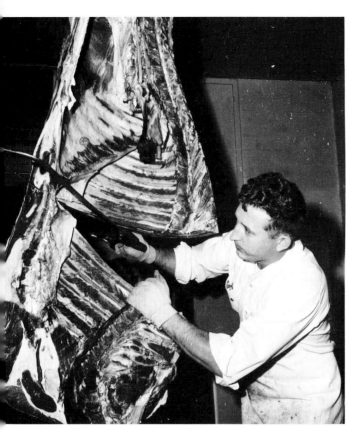

Breaking down the carcass into primal cuts is accomplished at the packer-wholesaler level. Circa 1965.

Interior of a meat packing plant in the 1950's showing carcasses hanging on the hook.

From a home on the range at Richardson Ranch near Hooper, Washington these steers could move to a large feedlot operation in the area owned and operated by the McGregors.

Holding pens at Marysville, Washington. Cattle are sorted, grouped for uniformity prior to being brought to the sales ring for auctioning.

No part of the beef business is more important than marketing. Sizing up a pen of cattle is a challenge to the buyer and means profit or loss to the seller. Cattle are sold by direct negotiation, private treaty between buyer and seller, or taken to marketing centers where they are auctioned off to the highest bidder. Livestock markets charge a flat fee per head for the service.

Courtesy Northwest Unit Farm Magazines

The Cariboo is known for its cattle from the ranges of the Chilcotin. Many of the animals are sold through the largest stockyard in the interior of British Columbia. Elmer Derrick, local manager for the British Columbia Livestock Producers Association, checks the stock as they move to the pens.

Courtesy Williams Lake Tribune.

The inside story. The man at the "mike" is the auctioneer, who, with rapid-fire chanting, calls for bids on the animals in the pen or "ring."

Three "ringmen" scan the audience for bids on the lot of cattle. Circa 1967. Bitter Root Stockmen's Association cattle, Montana.

Courtesy Burlington Northern

Aerial view of the Williams Lake Stockyards.

Courtesy Williams Lake Tribune

A view of the livestock market scene, center of marketing activity. It is located adjacent to railway and highway for convenient delivery prior to the sale and removal of cattle after. The heaviest marketings occur in the fall.

Portable scales facilitate checking weight at various stages of the animal's growth. Rate of gain is a guide to raising good beef.

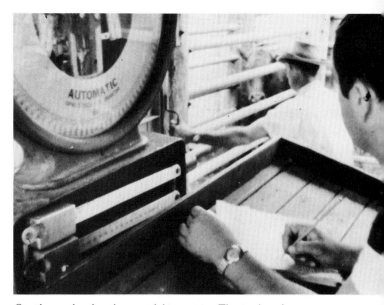

On the scales is where weight counts. The scales shown are stationary.

Dinner is served! Feeding on Elmer Schneidmiller's Ranch in 1952 at Post Falls, Idaho.

Washington Water Power photo.

Steak on the hoof for Western families. Montana beef ready for shipment to market.

Courtesy Burlington Northern

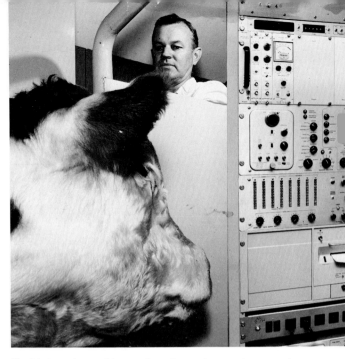

"Hedging" on cattle futures became a part of the cattle industry's future in the 1960's. These were the first animals to be delivered under terms of a Chicago Board of Trade contract in Choice steers.

Wanapum Dam
Huge dams on the Columbia River and its tributaries have brought thousands of acres under cultivation in the Northwest. Row-crop waste products such as sugar beet tops, can be utilized in feeding cattle.

Sophisticated machinery has been invented to replace "eyeballing" (visual determination) of the amount of fat and lean on an animal. A polled Hereford is being scanned in a walk-through type unit. It required about three minutes to measure lean meat to within seven and one-half pounds.

Harry Lindeman feedlot.

Courtesy of Northwest Unit Farm Magazines.

By 1942, the benefits of feeding or finishing cattle was well established. Gathering at the feed trough is a soft life compared to grazing the range. Jack Ray cattle at De Smet, Montana.

Courtesy Burlington Northern. Circa 1942.

Buffalo Rapids Project, Montana — Art Martin, Feed Boss for the Ruben Schroeber feed lot near Terry, Montana keeps the bunks full of silage and alfalfa pellets for the already contented creatures. The feedlot is supplied by Rubens irrigated farm under the Shirley Main Canal and by alfalfa pellets grown and processed in other irrigated areas of Eastern Montana. 7-26-66 Bureau of Reclamation photograph by Lyle C. Axthelm.

Courtesy of the Bureau of Reclamation
Department of the Interior

YESTERDAY

Improved methods in the care, breeding and feeding of cattle contribute to finer textured, more flavorful meat. Modern handling and storing methods keep beef fresh and clean at the packing plant and retail meat counter. There is a constant supply of high quality beef for selection by families of the Northwest.

The finished product, tempting beef for the table.

INDEX

Carnation Farms, 79
Carrie Ladd, 17
Cattle drives, 17, 21, 59, 60, 64, 65
Cavvy, 91
Celtic Ox (Bos longifrons), 9
Challis, Idaho, 55, 123
Charlie, Heneas, 95
Charolias Cattle, 161, 167
Cheney, Washington, 134
Cherry Creek, B.C., 40
Chianina Cattle, 163
Chicago, Illinois, 175, 184
Chicago Livestock Market, 176
Chuck Wagons, 71, 100, 101, 102, 103
Clarks Fork, Idaho, 141
Clausen, Alma Z., 173
Clinton Hotel, 13
Cloverland, Washington, 161
Coleman, W. L., 97
Cooke, C. P., 25
Colville Indian Reservation, 128
Colvin, Mrs. Tom, 161
Conrad Circle Company, 102
Cornwall, Mrs. Hugh, 26
Cotton Ranch Cowboys, (Chilcotin
    Indians), 97
Cotton Ranch, Williams Lake, B.C., 55
Cotton, R. C., 168
Cowboy Camps, 104, 106
Cowboys, 17, 68,-72, 84, 88, 90, 93-99,
    105, 119
Coulee City, Washington, 108
Cowiche River, Washington, 74
Circle Roundup, 95
Crawford, Alex, 63
Creston, B.C., 42
Crow Indian Reservation, Big Horn
    County, Montana, 91, 92
Crow Indian Cowboys, 150
Cruse, Thomas, 96
Curtice Martin Ranch, Montana, 162

— D —
Daily, Lloyd, 161
Daugherty, Oscar, 96
Davenport, Washington, 128
Davis, 29
Davison Brothers, 26
Deer Lodge, Montana, 30, 39
Deer Park Ranch, B.C., 33
Dement, Samuel, 145
Denby, Al, 69
Devine, John S., 29
Devon Cattle, 26, 160
Dexter Cattle, 167
Dexter, Jim, 119, 121

Dick Ranch, Asotin County,
    Washington, 173
Dick Ranch Cowboys, 99
Dillard, Tom, 68
Doan's Crossing Monument,
    Oklahoma, 61
Dooly, Clarence, 36, 58
Dorman, W. J., 123
Douglas Lake, B.C., 38
Douglas Lake Ranch, 130
Driggs, Idaho, 138
Drumheller, George, 31
Drumheller, Jesse, 26
Drumheller, Tom, 26, 31, 144
Drumheller, "Uncle Dan", 25
Duff, Roy, 146
Dunn, Dr., 168
Dupier, Len, 136
Durham Cattle, 16, 158, 161

— E —
Edgmand, Pete, 94
Edmonton, Alberta, Canada, 39
Edwards, Indian, 95
E and I Roper Farm, Victoria, B.C., 172
Ellis, Ervin, 77
Emmett, Idaho, 81

— F —
Fairfield, Idaho, 113
Fallon, Montana, 99
Father Larkin, 150
Fender, Porter C., 76
Fences, 125, 127
Ferrell, John, 16
"Figure 3" Homestead, 36
"Figure 2" Ranch, 152
"Figure 3" Ranch, 152
Fitzpatrick, Bob, 91
Flathead Valley, Montana, 10
Fleckvie Cattle, 163
Fleegle, Lee, 148
Fletcher, Orville, 98
Flowers, John, 69
Flynne, Jack, 104
Foaght, Bill, 148
Forest, Don, 153
Fort Ellis, Montana, 54
Fort Nisqually, Washington, 17
Fort Vancouver, 15, 17
Fowler, Stewart H., 60
Fraser River, B.C., 18, 64, 72
Frear, Fred, 177
Freeland, Frank, 69
Freight wagons, 19, 21